# Flexible and Wearable Electronics: Design and Fabrication Techniques

## Haider K. Raad

### Xavier University, USA

United Scholars
Publications

United Scholars Publications, USA
Our distinguished editorial team thrives to produce and publish
superior quality books and journals.
Visit our website for the current list of publications
www.unitedscholars.net

Editor: Haider K. Raad, Ph.D.
Department of Physics
Xavier Wearable Electronics Research Center (XWERC)
Xavier University, Cincinnati, Ohio, USA

Published by United Scholars Publications, USA
Copyright © 2016 United Scholars Publications
**www.unitedscholars.net**
**info@unitedscholars.net**

**ISBN-13:** 978-0692751718
**ISBN-10:** 0692751718

**Disclaimer**
The Publisher and the Editor\Author hold no liability for incidental or consequential injuries or damages caused by the information contained in this publication.

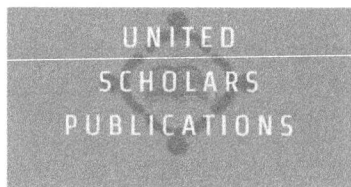

UNITED
SCHOLARS
PUBLICATIONS

# ABOUT THE EDITOR

Dr. Haider K. Raad currently serves as the director of the Engineering Physics program and the Wearable Electronics Research Center (XWERC) at Xavier University in Ohio, USA. He was previously affiliated with California State University and the University of Arkansas at Little Rock between 2008 and 2015. Dr. Raad teaches undergraduate and graduate level courses such as Electronic Circuits, Antenna Engineering, Electromagnetics, and Wireless Systems. Haider received the Ph.D. and M.S. degrees in Systems Engineering, specializing in RF Telecommunication systems from the University of Arkansas at Little Rock (UALR), and the M.S. degree in Electrical and Computer Engineering from New York Institute of Technology (NYIT). Haider is also the editor of the best-selling book entitled `Innovations in Wearable and Flexible Antennas` published by WIT Press, UK, and has recently published a book entitled 'Telemedicine: Emerging Technologies, Applications, and Impact on Health care Outcomes'. He has also published several book chapters, and over a hundred peer reviewed journal and conference papers on research fields of his interest which include: flexible and wearable wireless systems, Telemedicine and Wireless Body Area Networks, Metamaterials, MIMO, and biomedical electronics. He is also the recipient of the E-Telemed Best Paper Award, SSU's Research Fellowship Award, UALR's Outstanding Teaching Support Award, and AAMI/TEAMS Academic Excellence Award.

*To my beloved parents: Nawal and Raad, and precious sisters:*
*Tamara and Ramla*

# Preface

Currently, the field of flexible and wearable electronics is recognized as one of the hottest research topics in academia and research and development centers. Flexible and wearable electronics are also marching steadily toward becoming mainstream thanks to the fast advancements in the digital systems and electronics manufacturing and packaging. The number of wearable and flexible devices available in today's electronics market including fitness trackers, smart glasses and watches, and smart clothing seems to grow very rapidly.

Such devices, which are usually low-profile, light-weight, bendable, portable, and reconfigurable would considerably expand the applications of modern consumer electronic devices.

According to a recent study, the global revenue of this market is estimated to be 45 billion USD in 2016 and over 300 billion USD in 2028.

The availability of inexpensive flexible substrates in addition to their light weight, energy efficiency, reduced fabrication complexity and cost, make flexible and wearable electronics a reasonable and attractive alternative to the current technology which is based on rigid platforms. It is also worth mentioning that the recent advancement in miniaturizing an efficient and flexible energy storage components have served as an important catalyst for the success of this technology.

To the best of the Author's knowledge, there is no dedicated volume targeting the current research and development trends of flexible and wearable electronics; hence, I felt the necessity to compile an edited book on this vital subject.

The objective of this book is to provide a comprehensive guide to the state of the art technologies and methods applied in the realization of flexible and wearable electronics. The targeted readers of this book include but not limited to college professors, Research and Development scientists, practicing electronics and material engineers, in addition to all the enthusiasts interested in modern technological advancements. Moreover, the book serves as an extensive resource for graduate students working on topics related to wearable and flexible electronics.

This book, organized into eleven chapters, introduces the latest research findings and trends related to the design, fabrication processes and techniques used in the realization of wearable and flexible electronics.

**Chapter One** shows how recent developments in wireless, flexible, fully-printable, and sensor technologies are paving the way for the emergence of ubiquitous sensing capabilities, in the form of "Smart Skins" and the Internet of Things (IoT). The chapter also presents a brief review of state-of-the-art implementations of flexible chemical sensing components technologies; and reviews the latest advances in wireless interrogation techniques for such devices.

**Chapter Two** presents the incremental Latin Hypercube Sampling (LHS) technique which is aimed at speeding up the circuit simulations while considering process variation and aging effects. Two automated robust design optimization techniques are introduced for the analog circuits with flexible Thin Film Transistors.

**Chapter Three** presents the advantages and suitability of semiconducting Nano Wires (NWs) for flexible and large area electronics. Various aspects of NWs such as physical and chemical properties, synthesis methods, processing, and applications are reviewed with examples.

**Chapter Four** discusses how to fabricate flexible/stretchable piezoelectric devices by the use of electrohydrodynamical direct-writing technique. Furthermore, this chapter provides detailed engineering design rules and paves a cost-effective and high-efficiency manufacturing pathway for applications in wearable and bio-integrated electronics.

**Chapter Five** explores chief approaches towards the development of flexible silicon electronics. The chapter reports some of the most promising methodologies for converting rigid silicon substrates into flexible electronics through the use of silicon-based nanomaterials, etch-release processes, and stress-based wafer spalling. It also presents a thorough study of the most advanced transistor architecture (FinFET) and its integration within a flexible silicon platform.

**Chapter Six** focuses on the use of inkjet printing technique of conductive tracks and electrode patterns on textile substrates. The chapter also reports a two-step fabrication process using UV barrier channels and low viscosity functional inks.

**Chapter Seven** provides an overview of existing wearable technologies for Intra-Body Communication (IBC); it also addresses their potential and challenges, and discusses future directions.

**Chapter Eight** presents the design of a flexible antenna structure operating at multiple frequencies covering WLAN, WiMaX, GPS, and ISM applications.

**Chapter Nine** discusses the design of Near Field Communication (NFC) wearable antennas. The chapter examines the design of such antennas especially for small gadgets like wristbands, watches, and rings. Considerable focus has been given to different shapes, sizes, and materials for designing these antennas.

**Chapter Ten** reports recent advances in the fabrication, structure, mechanism and electrical output performances of various fiber-based and textile-based flexible generators, including piezoelectric, triboelectric and electret generators. Moreover, discussions are provided regarding newfangled applications, current challenges and future directions.

Last but not least, the appropriate choice of the electronic elements of wearable sensor nodes including the power source, sensors, digital signal processing components, and transceivers to build up an energy-efficient wearable body area network is discussed in **Chapter Eleven**.

Haider K. Raad, Ph.D.
Xavier University, Ohio (USA)
July 2016

# Table of Contents

*Chien-Nan Jimmy Liu, Yen-Lung Chen, and Nguyen Cao Qui*

**Chapter 3:** Inorganic Semiconducting Nanowires for Flexible
Electronics ...................................................................73
*D. Shakthivel, C. García Núñez, and R. Dahiya*

**Chapter Four:** Flexible/Stretchable Piezoelectric Nanofiber
Devices .....................................................................101
*Yongqing Duan, Yajing Ding, Youhua Wang, Yewang Su YongAn
Huang*

# CHAPTER ONE

## Flexible Fully-Printed Low-Cost Zero-Power Wireless Chemical Sensor Technologies for Ubiquitous  Monitoring

Jimmy G.D. Hester and Manos M. Tentzeris

*School of Electrical and Computer Engineering, Georgia Institute of Technology, Atlanta, Georgia, USA*

## 1. Introduction

Not often do emerging technologies offer the potential of having a truly transformative impact on societies and the lives of their individuals. The 20th century has witnessed such a phase with the birth, and exponential development, of information technologies, which now provide an unprecedented and low cost medium for information access and exchange to billions of individuals across the world. Nevertheless, this rich information realm can only be accessed through a handful of devices, such as computers, smartphones and tablets; a clear-cut boundary with limited interface points still separates physical and information worlds. However, innovators and forward thinking businesses, fueled by large early adopter communities, are driving a rapidly growing technological trend towards the interweaving of both universes, with technologies such as virtual reality, additive manufacturing technologies (such as 3D and inkjet printing), and ubiquitous (including wearable) electronics. In this two-way path of digital to physical (actuation) and real to digital (data collection) exchange, sensors play an essential role: that of transforming physical parameter measurements into store-able and processable digital data. In this chapter, we will show how recent developments in wireless, flexible, fully-printable, sensor technologies are paving the way for the emergence of ubiquitous sensing capabilities, in the form of "Smart Skins" and the Internet of Things (IoT). In the first part, we will show the substantial potential scope and importance of low-cost ubiquitous sensing, before demonstrating the unique appeal of printed flexible electronics towards attaining this capability. The second part will present a brief review of state-of-the-art implementations and demonstrate the broad range of flexible chemical sensing components technologies. Finally, the third part will review the latest advances in wireless interrogation techniques for such devices.

## 2.    Printed, Low-Cost, Flexible, Zero-Power Wireless Systems for Ubiquitous Monitoring and Sensing

Ubiquitous chemical monitoring would provide solutions to many modern technological, industrial and ecological challenges, which will be presented in this section. A quick analysis of the required capabilities of such devices will then be conducted to demonstrate the appeal of printed, low cost, flexible, zero-power wireless systems for these applications.

### 2.1  A Broad Range of Applications

### 2.1.1  Product Monitoring

The use of radio frequency identification (RFID) tags is already quite common in logistics and inventory tracking, where the technology answers the demand for quick and automated item identification. In addition, the integration of sensors in packages or containers for tilt, shock, temperature or humidity monitoring is also growing. Despite these trends, wireless-enabled sensor devices are quite rare, with their implementation being limited to battery-powered systems, whose cost remain prohibitive for widespread usage with limited-value products. Tracking the freshness of food and perishable liquids from the shipping facility to the inside of the consumers' refrigerators, monitoring the proper storage conditions of sensitive medical consumables, or the quasi-real-time checking for the mishandling of fragile packages are important challenges that an emerging generation of low cost wireless sensor nodes (a.k.a. motes) could meet.

### 2.1.2 Chemical and Biological Hazard Detection

Chemical and biological hazards can represent an important potential threat in a number of industrial or domestic environments; most notably in the chemical and fossil fuel industries. The sensors presented in this work could (quite literally) be used as the canaries in the coal mine, by not only providing a quick response system, thanks to their ubiquitous coverage, but also transmitting precise information as to the spatial localization of threats and, therefore, allowing its quicker neutralization. Such capabilities could also be implemented into warfare environments as well as large public spaces for the prevention and countering of biological and chemical weapons attacks.

### 2.1.3 Agricultural Optimization

The US Environmental Protection Agency estimates that 10000 to 20000 physician-diagnosed pesticide poisonings occur each year, in the United States [11]. This problem, along with the need for continuous and precise monitoring of the levels of crops pesticide exposure could be tackled by the use of low cost ubiquitous sensors. Furthermore, precise real-time knowledge of nutrients, potential biological and chemical contamination and humidity levels, especially in controlled green-house and/or hydroponic environments, would permit a much more cost-efficient, environmentally-friendly and growth-optimal use of fertilizers, water, and phytosanitary products.

### 2.1.4 Environmental Surveillance

Environmental agencies, such as the EPA strive to continuously monitor chemical and biological pollution levels in air, water, and land over large natural environments. Such monitoring usually relies on both regular sample collection followed by laboratory analyses and in-situ measurements, using complex base stations. Both of these methods are capable of providing reliable measurements, but can only cover a limited set of points of interest, and are rather expensive. The emergence of reliable sensing technologies in the model of those presented in this chapter would allow for the implementation of dense meshes of fully-autonomous wireless sensors networks for ubiquitous, low cost environmental monitoring.

## 2.2 Device Requirements

Ubiquitous sensing systems would, ideally, have a range of properties that would allow them to achieve broad and rapid market adoption. Three main capabilities can be initially considered.

### 2.2.1 Wireless Functionality

Motes, by definition, need to be interrogated regularly in order to extract the measured sensing information. With the high mote density envisioned for ubiquitous sensing, wired configurations would be quite costly and challenging in implementation, while inducing very low system flexibility. As a consequence, wireless capabilities are essential for such systems.

## 2.2.2  Energy Autonomy

A consequence of the requirement for wireless operation is the simultaneous necessity for the use of either entirely passive devices, capable of functioning without the need for an internal power source, or that of energy-harvesting or battery-enabled devices. Due to the maturity of current battery-powered wireless electronics, most of the currently commercialized sensor nodes are designed around that last powering scheme. However, even with, at best, multi-year battery lives, such implementations induce costly battery changing maintenance operations, as well as high battery-material waste generation. For these reasons, dense-widespread-network-compatible motes will, in order to make economic and environmental sense, need to function autonomously either without the use of a power source (a.k.a zero power), or by relying on ambient radio-frequency, vibrational or optical energy harvesting.

## 2.2.3  Printability and Large-Scale Manufacturability

Gartner, Inc. predicts that there will be 20.8 billion devices on the IoT by 2020, up from 3.8 billion in 2014 and 6.4 billion in 2016 [14]. Such an exponential growth in the number of devices (battery enabled ones, to a greater degree) could have tragic and long lasting environmental consequences due to both manufacturing-induced and disposal/recycling-generated material and energy wastes. Therefore, the use of eco-friendly (or even completely biodegradable [1]) materials and processes will become essential for the emergence of the technologies discussed in this chapter. Additive manufacturing technologies (AMT)—a subset of which are printing technologies—may offer an extremely low cost solution for this problem. Indeed, such methods are, by definition, purely additive and, therefore, do not generate much direct waste during fabrication; only the deposited amount of materials needs to be used. Furthermore, these methods are very compatible (unlike standard fabrication methods) with a number of flexible low-cost and bio-degradable substrates, such as paper, silk, and PLA, as well as with flexible substrates, whose use also provides the advantage of allowing extremely compact conformal packaging integration. Finally, such technologies are also compatible with radio frequency (RF) electronics, and have already been implemented for the fabrication of a wide variety of high-performance sensing and origami-reconfigurable radio frequency (RF) structures [19, 29, 54]. Fully-printed devices are therefore exceptionally promising, thanks to their projected low cost, low fabrication and disposal environmental impact, as well as flexibility and large-scale manufacturability.

## 3.    Fully-Printed    Chemical/Biological    Sensor Technologies for Low-Cost Flexible Motes

This chapter will focus on giving an overview of the state of the art in printed or printable sensor technologies, used for the detection and concentration tracking of a range of chemical agents of significant practical importance. Due to the broad extent of their current and potential application contexts in food and perishable liquid monitoring, occupational hazard detection, and environmental surveillance, the development of chemical and biological sensors holds a particular importance in sensor technology research. Two major types of sensors can be distinguished: gas, and liquid-phase sensors. This chapter will only cover electrical component-based technologies that are compatible with flexible substrate. Therefore, optical characterization techniques which require complex and power-hungry interfacing electronics, or piezoelectric-material-based techniques, such as quartz crystal microbalance (QCM) and surface acoustic waves (SAW) methods, that require the use of non-flexible piezoelectric substrates will not be covered. Flexible-substrate-compatible sensors usually function according to two main principles, whose implementations generally are that shown in Fig. 1.

- Amperometric and potentiometric sensors fundamentally rely on either the current induced by an ingoing chemical reaction with the targeted analyte [48], or the measured equilibrium potential of an electrochemical cell [12], respectively. These sensors already have a widespread use in a number of industries; many glucose or pH sensors in the market, as well as the majority of automotive exhaust fumes analysis equipment rely on the use of such technology. Both techniques rely fundamentally on the electrical behavior of an electrochemical cell and therefore require the use of electrolytes. This characteristic has major consequences as to these sensors ease of fabrication for low cost printed flexible sensor systems, for gas or liquid-phase detection. Such implications will be expanded upon in the next subsection.

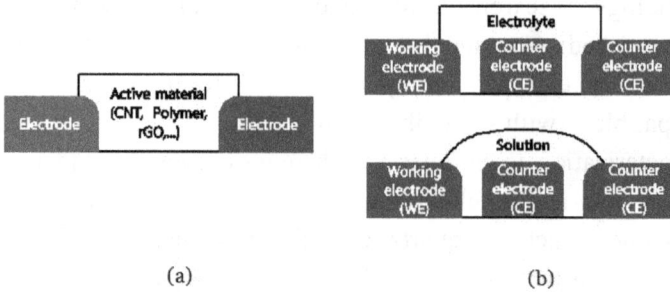

(a)                                    (b)

Figure 1: Cross section structure of typical (a) resistormetric and technologies    (b)    structural    or    solution    electrolyte potentiometric/amperometric sensors

- Resistometric/conductometric/impedometric sensors rely on a measured change of resistivity (or complex impedance) of the sensing component, as a consequence of the exposure to the gas analyte. While some amperometric sensors are often considered resistometric, as the measurement of a current, at fixed bias voltage, can be considered to be equivalent to a change in conductivity, the underlying mechanism is quite different. Indeed, the observed change in conductivity of an impedometric sensor is the product of variations of intrinsic physical properties (electronic energy levels, Schottky barrier heights [18], intrinsic capacitance [39, 46]) of the conducting, dielectric and semiconducting materials and/or structures ; not that of carriers added to the system as a product of redox chemical reactions. A major consequence of this, at great interest here, is the lack of requirement for the use of electrolytic materials. These components generally have an extremely simple structure that is comprised of two conductor (generally metal) electrodes connected by an active-material. Impedometric (mostly resistometric) sensors have a rather long history, with the first proof-of-concept demonstrations dating back to Bell labs in the 1950s [5]. However, the modern rise of nanotechnologies has breathed a new life into this sensor technology, especially for their use in low cost flexible printed systems, for three main reasons:

a- The fundamental importance of a high active material surface area for an enhanced performance was quickly recognized. This was later implemented, not only through the use of semiconductor nano-crystallites or nano-wires with, indeed, significant performance improvement [57], but mostly with semiconducting and conducting 2D nanomaterials and conjugated molecules, which gave rise to a range of new implementations with extremely high surface area and unprecedented performance.

b- New nanomaterials, such as graphene, carbon nanotubes, and conjugated molecules and polymers can very easily be printed on virtually any materials, without requiring any subsequent high temperature sintering.

c- These new materials greatly expanded the chemical and physical diversity of the active-gas-sensing-materials spectrum, allowing for the effective detection of a greater range of analytes.

Another important class of this broad impedometric sensor family is that of permittivity-tunable (or, conceivably, permeability-tunable) RF structures, where the entire RF device acts as both the sensor and the tag, taking advantage of electromagnetic principles [20]. For example, polyimide, used as a RF substrate, can provide an easily detectable resonance frequency detuning, thanks to its relative-humidity-dependent dielectric permittivity.

Both of the above technologies have been implemented for the (sometimes fully-printed) fabrication of a broad range of chemical sensors. Gas-phase and liquid-phase sensors will be discussed separately, as they provide different opportunities and challenges, and are generally results of the efforts of distinct research communities.

## 3.1 Gas Sensors

Gas sensors have the ability to track the concentration and/or detect the presence of specific gas-phase chemical agents present in their environment. Solid-state potentiometric and amperometeric gas sensors, have a number of properties that make them of challenging use for the purposes described in this chapter. Indeed, these types of structures require electrolytes for gas sensing, which can either be in solution or solid form. Solution or gel electrolytes are quite challenging from a printing and packaging perspective, while most of solid electrolytes require the use of highly crystalline films which, even though their material can be deposited using printing methods [7], require firing at temperatures in excess of 1000 °C, and generally require high operating temperatures, above 150 °C ; such heating (unless in a high-temperature environment) would increase the power consumption of these sensors to unacceptably high levels for the configurations considered in the present context. Nevertheless, viable printable electrolytic materials are slowly coming into play, with efforts reporting the use of carefully-packaged ionic gels or liquids [49], as well as plain or doped polymer matrices [55], which can be printed even on flexible substrates [60]. However, not much work has yet been done on the integration of such gas sensors into printed far-field wireless systems [25].

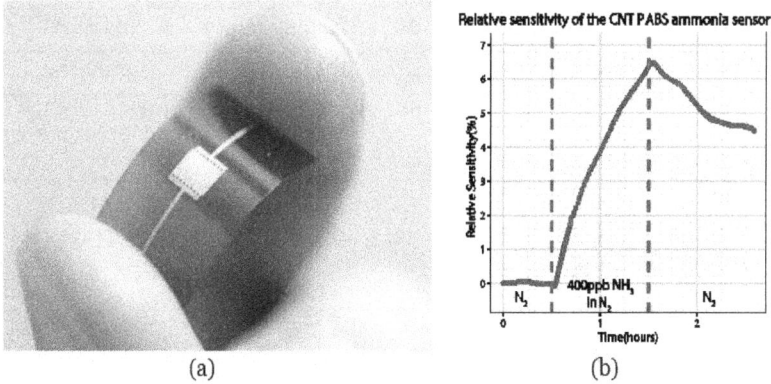

(a)                                                    (b)

Figure 2: (a) Picture of a fully inkjet-printed CNT-PABS-based resistometric flexible ammonia sensor and (b) its response to exposure to 400 ppb of ammonia.

Resistometric and impedometric sensors, whose fabrication is not constrained by the necessary use of electrolytic materials, have, on the other hand, been implemented (and characterized at DC [21] and RF [22]) in many occasions as fully-printable flexible vapor sensors, usable at ambient temperatures. Indeed, the properties previously mentioned have enabled the design and fabrication of high performance sensors for a wide range of gases, a subset of which are listed in Table 1. An example of a SWCNT-PABS ultra-sensitive ammonia sensor, fully inkjet-printed on polyimide substrate, and its response to 400 ppb of ammonia, is shown in Fig. 2.

Table 1: List of notable resistometric gas sensors, and their sensing materials

| Ref. | Analyte | Lowest detected concentration and associated sensitivity | Sensing materials |
|------|---------|----------------------------------------------------------|-------------------|
| [45] | H2 | 30% at 3 ppm | SiNW/Pd |
| [59] | CO2 | 10% at 10 ppm | Graphene |
| [24] | NH3 | 60% at 10 ppb | PANI-PSS NP |
| [10] | NO2 | 0.5% at 500 ppb | rGO |
| [40] | NO2 | 5% at 100 ppt | CNT-PEI |
| [4] | CO | 2.2% at 10 ppm | MWCNT |
| [33] | CH4 | 1% at 6 ppm | SWCNT-Pd |
| [37] | DMMP | 150% at 1 ppb | SWCNT |
| [31] | Nitrotoluene | 1.5% at 54 ppm | SWCNT |

The sensors listed here were mostly manufactured with different sensing materials, which enabled each one of these devices with particularly high sensitivity to their associated analyte. Nevertheless, most of those sensors are (to a certain degree) sensitive to many other chemical compounds. This lack of chemical selectivity is one the main challenges for this technology. Nevertheless, such a problem can be addressed by assembling several of these different sensors into an array [34, 35, 47], and taking advantage of their collective response to the same chemical excitation; a properly chosen set of sensors can then provide a response that can be uniquely associated to a given chemical environment (in concentration and composition).

## 3.2  Liquid-Phase Sensors

The conditions of liquid phase sensing are quite different, when considering electrochemical sensors. Indeed, the presence of a liquid bridging the electrodes of the sensor cell removes the necessary use of printable room-temperature-operating electrolytic bridge. As a consequence, the sensor structure is effectively reduced to mere electrodes, which can then be functionalized in order to sense the targeted analytes. Given the additional appeal of the low cost of printing methods, it is therefore no surprise that the literature abounds of reports of such sensors [56]. In addition, the range of demonstrated detectable analytes is extensive, both in the areas of biological [23] and chemical sensing. The area of wireless liquid-phase sensing nodes has not drawn as much attention as gas sensing applications have. One of the reasons for this is the relative incompatibility between liquids (especially water) and RF structures. Indeed, a radiating structure immersed in water, over the frequency range compatible with maximum practical dimensions for long-range radiating wireless sensor nodes, is electromagnetically isolated from its environment. Nevertheless, interfacial configurations (such as floating nodes) or and in-air moisture absorbent devices could very well be considered.

## 4.  Wireless Interrogation Technologies for Zero-Power RFID and Wireless Sensing Motes

As previously argued, it is essential for these motes to be enabled with wireless capabilities. Given the parallel requirement for a low (or zero) power operation, two broad classes of systems can be considered: on the one hand, low power active electronics (which are not yet printable) can be used in order to build energy-harvesting-powered systems with extremely low power consumption while, on the other hand, fully-passive chip-less systems offer zero-power consumption and lower complexity, but with limited capabilities.

## 4.1  Low Power Circuitry Implementations

Low power circuitry implementations are typically active systems and, as a consequence, require internal power sources. Nevertheless, their low power consumption allows them to function even on the limited power that they are able to scavenge in their environment. One architecture, that of Radio Frequency IDentification (RFID) tags has achieved widespread commercial use initially for other purposes than sensing, but has, as a consequence of its success, attracted much work in attempts to enable it with sensing capabilities. On the other hand, several very promising custom low-power backscattering systems have also been proposed.

### 4.1.1  Typical Passive RFID Systems

RFID systems can either be passive, semi-passive or active. These different declinations correspond to powering and communication scheme combinations. Active tags require an internal power source (a battery, usually) and use this power to generate RF waves for communication, as well as for running other operations (sensing, data collection and processing, etc.). Semi-passive tags are also battery-powered but rely on backscattering of the impinging reader-generated wave for communications; other processes are dependent upon the battery. Finally, passive tags also use backscattering for communication, but also rely solely on harvesting of the impinging reader wave to power their operation. For the reasons detailed in Sec. 2.2.2, only this last category will be considered here. Furthermore, because of the focus on long range applications and the extremely small range (typically less than 1 m) of lower frequency configurations, only UHF (around 900 MHz) implementations will be discussed. The literature offers numerous examples of the use of off-the-shelf UHF passive RFID chips for sensing purposes, ranging from humidity sensors [13, 52] to ammonia [38, 58] and touch [26] sensors.

A passive tag is constituted of an RFID chip, and an antenna. In the first phase of the reading cycle, the reader sends a continuous wave towards the tag, which will then be harvested in order to power the chip. The minimum amount of power (radiated by the reader antenna) required to turn the tag on can be calculated by [36]:

$$P_{max} = \frac{4\pi R^2 P_{th}}{\lambda G_t G \rho \tau} \qquad (1)$$

where $\lambda$, R, Pth, Gt, G, $\rho$, $\tau$ are, respectively, the wavelength, the reading distance, the chip power threshold sensitivity, the gain of the reader antenna, the gain of the tag antenna, the polarization efficiency, and the impedance matching coefficient between the tag chip and antenna. By being integrated into the radiating or matching structure, sensing components/films can, here, be used to induce changes in $\tau$ upon sensing. Therefore, for a known distance and configuration (which can be determined by the use of a reference tag [26]), a remote determination of the lowest reader-emitted power required in order to receive a response from the tag indirectly gives $\tau$, which can then be related to the properties of the sensor, and, finally, to the sensed parameter. As previously mentioned, the advantage of this mechanism is that it can be used with off-the-shelf, high-performance passive RFID chips. However, there are a few drawbacks that need to be resolved before a widespread implementation can take place. Indeed, the integration of sensor elements is not necessarily straightforward. For example, high resistance (more than 1 kΩ) resistometric sensors cannot effectively be integrated into the structure (without the addition of active components) without greatly negatively impacting the reading range. Furthermore, the reading range of these systems (similarly to all other systems that exclusively rely on instant-power RF energy harvesting) is limited by power regulations. Indeed, the maximum reading range can be expressed as:

$$R_{max} = \sqrt{\frac{\lambda P_{EIRP} G \rho \tau}{4 \pi P_{th}}} \qquad (2)$$

where PEIRP is the maximum equivalent isotropic radiated power allowed by regulations. With state-of-the-art RFID chips, and current regulation values for these parameters, reading ranges are confined, in optimal cases, to values below 12 m.

### 4.1.2  Typical Low-Power Backscattering Systems

Unlike the systems presented in the last section, the devices described here are not commonly built around the typical off-the-shelf passive RFID chip architecture. However, because of their power-efficiency, backscattered communication schemes can easily be used in those architectures. Their typical structure is shown in Fig. 3, and includes an active circuit, one or several sensors, an antenna, a backscattering front end, and one or multiple energy harvesters (such a solar cell and a RF harvesting circuit). The most basic systems of this type are typically built around low power oscillators that are used to simply modulate in amplitude the signal during backscattering. This subcarrier frequency can then be received by the reader, and be correlated to a sensing component response and therefore encode the sensing data. Such a system prototype has already been implemented on a partially printed structure for ammonia sensing [32].

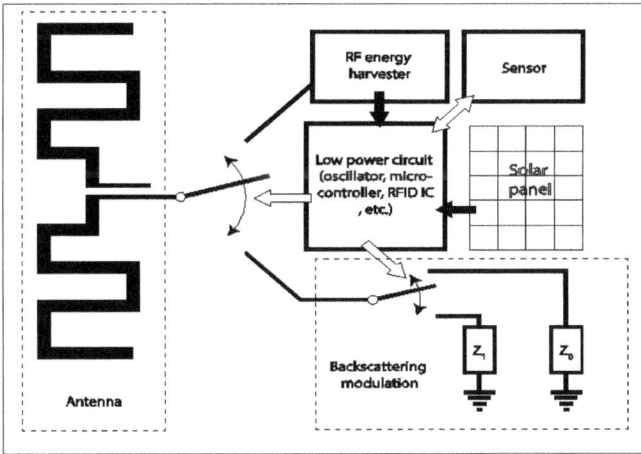

Figure 3: Schematic of a typical backscattering-communication low-power circuitry sensor structure.

The Wireless identification and sensing platform (WISP), first introduced in [43], has been popularized by its compatibility with RFID EPC protocols and its flexibility, as a general purpose architecture. Its architecture is comprised of an RF energy harvester, powering a low-power microcontroller which conducts sensor measurements and controls communications, through backscattering modulation. Nevertheless, this system being essentially a passive RFID on a board, with an added sensor interfacing capability [30], is more power consuming than highly optimized off-the-shelf passive RFID tags. As a consequence, its equivalent Pth (with reference to Eq. (2)) is higher, inducing a shorter maximum theoretical reading range of less than 2.5 m. In parallel, integrated sensing-enabled passive RFID systems have also been developed, such as in [6], where both temperature and light sensors were integrated into a passive RFID architecture. Commercial versions of these chips with internal temperature sensor and external sensor pins, such as the SL900A, from AMS, are now available. Nevertheless, they also suffer from the same cause for range limitation as that previously mentioned, and can only be read at maximum theoretical ranges of up to 3.8 m. As a consequence of the fundamental limitations of RF energy harvesting for the powering of these devices, subsequent efforts have focused on higher energy density sources, such as solar [41, 42] or multiform combinations of several harvesting sources [8]. The range of the WISP configuration was, in such a way, extended by a 6-fold, to 24 m [42]. Even though the use of currently used photovoltaic components can dramatically increase the cost of such devices, current trends in the development and cost-reduction of printed organic solar cells might bring commercial viability to such systems [9, 27].

## 4.2 Chip-less Technologies

As demonstrated in the last section, chips and backscattering communication enabled systems can provide autonomous solutions for wireless sensing, at ranges up to 4.5 m and 25 m by relying exclusively on RF and solar energy harvesting, respectively. Nevertheless, such systems rely on quite high performance electronic chips and components, which yet prevent them from being fully printed, and tremendously increase their unit cost. Indeed, even though very rudimentary fully-printed RFID systems have been demonstrated at HF (13.5 MHz), limitations in the mobilities of printed semiconductor materials, as well as repeatability issues, and limitations in the minimum length of printable transistor channels are still preventing the emergence of fully-printable UHF RFID tags [50]. As a consequence, much attention has been given to structures that are usually referred to as chip-less RFID's. As hinted by their name, such tags do not require chips in order to operate, and therefore solely rely on their passive interaction with, and reflection of, impinging reader electromagnetic waves. As a consequence, such systems are passive and, therefore, do not require any power source: they are "zero-power". Furthermore, due to their lack of active components, such devices are generally fully-printable, and extremely low cost. However, their electrical structure is therefore time invariant (if not for the presence of environment-modulated sensing elements). Hence, no modulation can be applied on a signal for communication purposes, and different schemes need to be utilized for information encoding. These schemes can generally be classified into two methods: frequency and time-domain-based.

### 4.2.1  Ultra-Wide Band (UWB) Time-Domain Interrogation Approaches

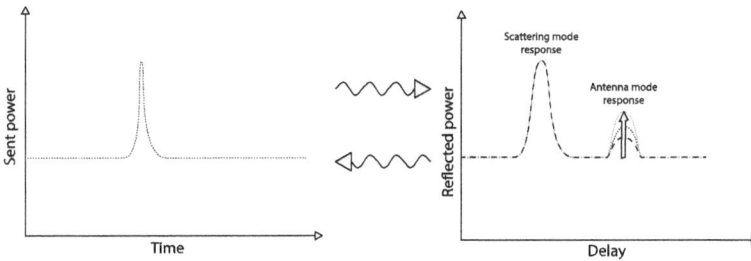

Figure 4: Principle of the UWB time-domain wireless sensor interrogation process.

The main challenge to achieve an acceptable detectability of wireless RF sensing components lies in the interference from the environment. Indeed, a signal emitted from a wireless reader will propagate in many directions, being reflected by the environment. Subsequently, the signal measured by the reader is a combination of several components that include noise, interfering reflections from the environment, interference from nearby RF sources, an interfering reflection from the antenna's structure (the structural mode), and an information-enclosing (the signal) antenna mode. The physical difference between those two modes can be partially understood by considering a standard patch antenna. At any frequency, a metal patch can be considered a metal plate which, as a metallic object, reflects electromagnetic waves in a way that is controlled only by its structural shape. However, at its resonance frequency (and within its bandwidth), the structure also behaves as an antenna, absorbing a part of the signal (in its antenna mode), which can then be re-emitted. Refer to [16, 17] for a more detailed and rigorous description. A major challenge is, therefore, to isolate, in reception, the information-enclosing antenna mode signal from the interference. A standard modulator makes this process quite straightforward by modulating the antenna mode, and therefore allowing a differential measurement to be conducted. Chip-less time-domain methods, which cannot rely on modulation, instead isolate different modes and interference by taking advantage of the (time-domain) phase delay, as seen in Fig. 4. The encoding of data is achieved by connecting the tag antenna to a delay line that is terminated with a load whose impedance changes upon sensing. In the most common reading method using UWB readers, a short pulse is sent, and the response is time-gated, allowing for a precise detection of the variations in phase and magnitude of the received reflected antenna-mode wave, which can then be correlated to changes in sensor impedance and, therefore, in sensed parameters. Another commonly used interrogation method uses a frequency modulated continuous-wave (FMCW) radar system whose spectral output can be correlated to the delay and magnitude of the reflected signal. UWB and FMCW systems have been demonstrated for ethylene [44], pressure [3], and temperature [15] sensing, but could be easily extended to many other chemical and physical parameters, using the sensors described in Sec. 3.

The main advantage of this technology is its ability to extract, even in some practical environments, very precise magnitude and phase information associated with the sensor antenna mode. Nevertheless, such readings are theoretically very sensitive to interference from objects at similar ranges (and therefore delay) as the sensor. As a consequence, even though those sensors are generally demonstrated in echoic environments, special attention is paid to place them away from nearby interferers (items are never placed on a wall, for instance). Furthermore, because of the short time-pulses that are necessary for UWB applications, the required absolute bandwidths are generally in order of more than 3 GHz. At the relatively low operation frequencies of the demonstrated prototypes (usually centered around 3 GHz), this represents a very wide relative bandwidth, which occupies much of the spectrum.

The use of these devices at higher frequencies could relax these UWB hardware and spectral occupation constraints, but this offers challenges of its own, that will be discussed in the last section.

## 4.2.2    Frequency-Domain    Resonance    Interrogation Approaches

In contrast with the time-domain-based technologies described in Sec. 4.2.1, frequency-based sensing relies on spectral resonances to encode information; the reader scans through the frequency band, and measures the magnitude of the reflected signal at different frequencies. Because of its ease of use and manufacturing, this technology has attracted much attention, and has been utilized for carbon dioxide [51], and humidity [2] sensing. Nevertheless, it cannot easily differentiate between the useful antenna-mode signal response and that of the interference signals without proper use of gating.

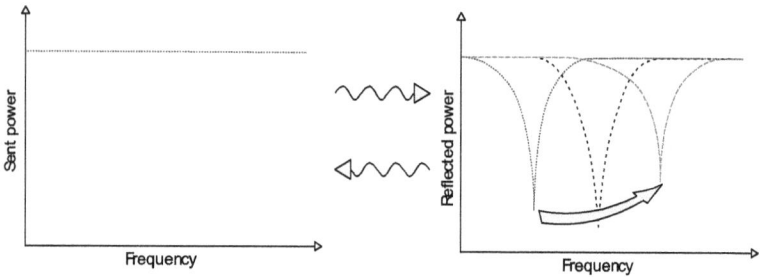

Figure 5:   Principle of the typical frequency-domain-based wireless sensor interrogation process

A clean signal, as the one shown in Fig. 5, is typical of measurements in anechoic environments. While it is possible to gate the received signal, using precise enough gating to isolate the useful signal (especially from the device scattering mode) would require, similarly to the time-domain-based method, the use of large frequency bands. As a consequence, these devices are consistently demonstrated either at up to 2 m ranges in anechoic chambers, or at tens of centimeters, away from possible interferers, in practical environments. Recent improvements have, nevertheless, been demonstrated with the use of cross-polarizing structures [28].

### 4.2.3 Van-Atta Reflectarrays

As previously mentioned, the biggest limitation in frequency-based chip-less systems is their limited capability to isolate the information-carrying signal from environment and scattering-mode-mediated interferences. Three general strategies can be utilized to enable and enhance the extraction and isolation of the sensing-information-embedded antenna-mode signal:

- Increasing the relative amplitude of the useful (antenna) signal with reference to interference and noise: One way to do this is to shift towards higher operation frequencies. Indeed, given fixed system (reader and interrogated sensor node) dimension, operation at higher frequencies leads to a higher detectability and, therefore, extended reading ranges. The power of the antenna-mode signal received by a reader can be expressed as

$$P_r = P_r e_{a,r} e_{a,t} A_r A_t F_t(\Theta) \left(\frac{f}{c}\right)^4 \quad (3)$$

where f, c, R, Pe, Pr, ea,r, ea,t, Ar, At and Ft($\theta$) are, respectively, the frequency, the velocity of light, the reading range, the emitted power, the received power, the aperture efficiencies of the reader antennas (assuming identical antennas), the aperture efficiency of the tag/target/sensor antenna, the physical aperture of the reader antennas, the aperture of the tag, and the radiation pattern of the tag. For fixed dimensions, signal power increases with f4. In addition, for a given reader antenna area, the directivity increases, and enables (by aiming) the isolation of environmental interference in directions away from reader antenna boresight. However, as the operation frequency increases, the tag radiation pattern ($Ft(\theta)$) also becomes much more directional, therefore limiting the range of readable directions to a small angular range around the tag antenna boresight.

- Introducing polarization differences (e.g. exploiting cross-polarization) between the useful signal and the interference, and enabling the reading system with polarimetric capabilities [53].
- Enabling the sensors and reading systems with absolute broadband capabilities, in order to enable a precise time-gating capability.

With those options in mind, one unique configuration stands out as a very appealing solution: The Van-Atta array [20]. Indeed, this fully-passive resonating structure, thanks to its coherent reflection capabilities, always reemits electromagnetic waves in the same direction as the impinging reader wave and, therefore, towards the reader. As a consequence, regardless of its size (and therefore detectability), such a structure can effectively be interrogated from any angle; $Ft(\theta)$ is virtually isotropic. Furthermore, such a structure can be designed to provide a cross-polarized response signal, with reference to the impinging reader wave and, therefore, to most of the environment and scattering mode interference. Very recently, such an approach was implemented with a fully-printed structure (shown on Fig. 6) for humidity sensing, and demonstrated in an unprecedentedly challenging real-life environment, at a record range for planar passive sensing structures of 5.5 m [20]. Measurements at ranges in extent of 30 m have been taken, and will soon be reported.

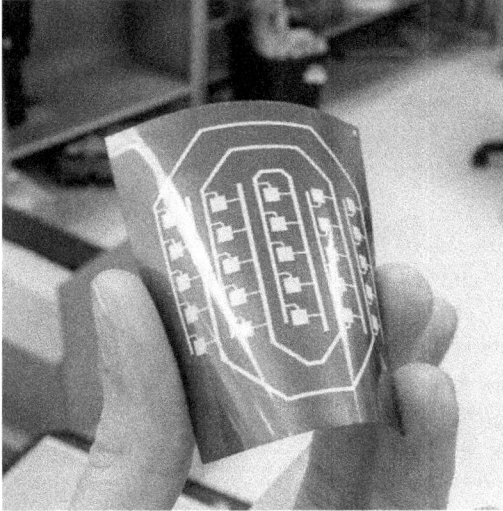

Figure 6: Picture of the printed Van Atta structure prototype.

## 5. Conclusion

The emergence of ubiquitous chemical sensors could have a deep impact in several industries and provide a safer and more environmentally preserved environment for future generations to inherit. Amongst potential technological solutions to meet this challenge, flexible printed wireless chemical sensors seem to offer an excellent potential, thanks to their capabilities, low cost and environmental-friendly footprint. Continuous advances in both printed sensor technologies and in low power printed (additively manufactured) wireless sensing systems are dramatically improving the capabilities and performance of such devices towards meeting the requirements of a growing number of possible applications, such as Internet of Things, smart skins and wearable biomonitoring.

# References

[1] Handan Acar, Simge C, ınar, Mahendra Thunga, Michael R Kessler, Nastaran Hashemi, and Reza Montazami. Study of physically transient insulating materials as a potential platform for transient electronics and bioelectronics. Advanced Functional Materials, 24(26):4135–4143, 2014.

[2] Emran Md Amin, Md Shoaib Bhuiyan, Nemai C Karmakar, and Bjorn Winther-Jensen. Development of a low cost printable chipless RFID humidity sensor. Sensors Journal, IEEE, 14(1):140–149, 2014.

[3] Herve Aubert, Franck Chebila, Mohamed Mehdi Jatlaoui, Trang Thai, Hamida Hallil, Anya Traille, Sofiene Bouaziz, Ayoub Rifai, Patrick Pons, Philippe Menini, et al. Wireless sensing and identification of passive electromagnetic sensors based on millimetre-wave FMCW RADAR. In IEEE RFID Technology & Applications, page 5p, 2012.

[4] C Bittencourt, A Felten, EH Espinosa, R Ionescu, E Llobet, X Correig, and J-J Pireaux. WO 3 films modified with functionalised multi-wall carbon nanotubes: morphological, compositional and gas response studies. Sensors and Actuators B: Chemical, 115(1):33–41, 2006.

[5] Walter H Brattain and John Bardeen. Surface properties of germanium. Bell System Technical Journal, 32(1):1–41, 1953.

[6] Namjun Cho, Seong-Jun Song, Sunyoung Kim, Shiho Kim, and HoiJun Yoo. A 5.1-µw UHF RFID tag chip integrated with sensors for wireless environmental monitoring. In Solid-State Circuits Conference, 2005. ESSCIRC 2005. Proceedings of the 31st European, pages 279–282. IEEE, 2005.

[7] WF Chu, V Leonhard, H Erdmann, and M Ilgenstein. Thick-film chemical sensors. Sensors and Actuators B: Chemical, 4(3-4):321–324, 1991.

[8] Ana Collado and Anthimos Georgiadis. Conformal hybrid solar and electromagnetic (em) energy harvesting rectenna. Circuits and Systems I: Regular Papers, IEEE Transactions on, 60(8):2225–2234, 2013.

[9] Seth B Darling and Fengqi You. The case for organic photovoltaics. Rsc Advances, 3(39):17633–17648, 2013.

[10] Vineet Dua, Sumedh P Surwade, Srikanth Ammu, Srikanth Rao Agnihotra, Sujit Jain, Kyle E Roberts, Sungjin Park, Rodney S Ruoff, and Sanjeev K Manohar. All-organic vapor sensor using inkjet-printed reduced graphene oxide. Angewandte Chemie International Edition, 49(12):2154–2157, 2010.

[11] EPA. Regulatory impact analysis of worker protection standard for agricultural pesticides. Environmental Protection Agency., 1992.

[12] J Fouletier. Gas analysis with potentiometric sensors. a review. Sensors and Actuators, 3:295–314, 1983.

[13] Jinlan Gao, Johan Siden, H Nilsson, and Mikael Gulliksson. Printed humidity sensor with memory functionality for passive RFID tags. Sensors Journal, IEEE, 13(5):1824–1834, 2013.

[14] inc Gartner. Gartner says 6.4 billion connected "things" will be in use in 2016, up 30 percent from 2015., November 2015.

[15] David Girbau, Angel Ramos, Antonio Lazaro, Sergi Rima, and Ramo´n Villarino. Passive wireless temperature sensor based on time-coded UWB chipless RFID tags. Microwave Theory and Techniques, IEEE Transactions on, 60(11):3623–3632, 2012.

[16] Robert Blair Green. The general theory of antenna scattering. PhD thesis, The Ohio State University, 1963.

[17] RC Hansen. Relationships between antennas as scatterers and as radiators. Proceedings of the IEEE, 77(5):659–662, 1989.

[18] S Heinze, J Tersoff, R Martel, V Derycke, J Appenzeller, and Ph Avouris. Carbon nanotubes as schottky barrier transistors. Physical Review Letters, 89(10):106801, 2002.

[19] Jimmy G Hester, Sangkil Kim, Jo Bito, Taoran Le, John Kimionis, Daniel Revier, Christy Saintsing, Wenjing Su, Bijan Tehrani, Anya Traille, et al. Additively manufactured nanotechnology and origami-enabled flexible microwave electronics. Proceedings of the IEEE, 103(4):583–606, 2015.

[20] Jimmy GD Hester and Manos M. Tentzeris. Inkjet-printed van-atta reflectarray sensors: A new paradigm for long-range chipless low cost ubiquitous smart skin sensors of the internet of things. In Microwave Symposium (IMS), 2016 IEEE MTT-S International, pages 1–4. IEEE, 2016.

[21] Jimmy GD Hester, Manos M Tentzeris, and Yunnan Fang. Inkjet-

printed, flexible, high performance, carbon nanomaterial based sensors for ammonia and DMMP gas detection. In Microwave Conference (EuMC), 2015 European, pages 857–860. IEEE, 2015.

[22] Jimmy GD Hester, Manos M Tentzeris, and Yunnan Fang. UHF lumped element model of a fully-inkjet-printed single-wall-carbon-nanotubebased inter-digitated electrodes breath sensor. In IEEE 2016 Antennas and Propagation Society International Symposium (APSURSI). IEEE, 2016.

[23] Christopher B Jacobs, M Jennifer Peairs, and B Jill Venton. Review: Carbon nanotube based electrochemical sensors for biomolecules. Analytica Chimica Acta, 662(2):105–127, 2010.

[24] Jyongsik Jang, Jungseok Ha, and Joonhyuk Cho. Fabrication of water-dispersible polyaniline-poly (4-styrenesulfonate) nanoparticles for inkjet-printed chemical-sensor applications. Advanced Materials, 19(13):1772–1775, 2007.

[25] Younsu Jung, Hyejin Park, Jin-Ah Park, Jinsoo Noh, Yunchang Choi, Minhoon Jung, Kyunghwan Jung, Myungho Pyo, Kevin Chen, Ali Javey, et al. Fully printed flexible and disposable wireless cyclic voltammetry tag. Scientific reports, 5, 2015.

[26] Sangkil Kim, Yuki Kawahara, Anthimos Georgiadis, Ana Collado, and Manos M Tentzeris. Low-cost inkjet-printed fully passive RFID tags using metamaterial-inspired antennas for capacitive sensing applications. In Microwave Symposium Digest (IMS), 2013 IEEE MTT-S International, pages 1–4. IEEE, 2013.

[27] Frederik C Krebs. Fabrication and processing of polymer solar cells: a review of printing and coating techniques. Solar energy materials and solar cells, 93(4):394–412, 2009.

[28] Bernd Kubina, Christian Mandel, Martin Schussler, Mohsen Sazegar, and Rolf Jakoby. A wireless chipless temperature sensor utilizing an orthogonal polarized backscatter scheme. In Microwave Conference (EuMC), 2012 42nd European, pages 61–64. IEEE, 2012.

[29] Vasileios Lakafosis, Amin Rida, Rushi Vyas, Li Yang, Symeon Nikolaou, and Manos M Tentzeris. Progress towards the first wireless sensor networks consisting of inkjet-printed, paper-based RFID-enabled sensor tags. Proceedings of the IEEE, 98(9):1601–1609, 2010.

[30] Taoran Le, Vasileios Lakafosis, Ziyin Lin, CP Wong, and MM

Tentzeris. Inkjet-printed graphene-based wireless gas sensor modules. In Electronic Components and Technology Conference (ECTC), 2012 IEEE 62nd, pages 1003–1008. IEEE, 2012.

[31] Jing Li, Yijiang Lu, Qi Ye, Martin Cinke, Jie Han, and M Meyyappan. Carbon nanotube sensors for gas and organic vapor detection. Nano letters, 3(7): 929–933, 2003.

[32] Yunfeng Ling, Hongtao Zhang, Guiru Gu, Xuejun Lu, Vijaya Kayastha, Carissa S Jones, Wu-Sheng Shih, and Daniel C Janzen. A printable CNT-based FM passive wireless sensor tag on a flexible substrate with enhanced sensitivity. Sensors Journal, IEEE, 14(4):1193–1197, 2014.

[33] Yijiang Lu, Jing Li, Jie Han, H-T Ng, Christie Binder, Christina Partridge, and M Meyyappan. Room temperature methane detection using palladium loaded single-walled carbon nanotube sensors. Chemical Physics Letters, 391(4):344–348, 2004.

[34] Yijiang Lu, Christina Partridge, M Meyyappan, and Jing Li. A carbon nanotube sensor array for sensitive gas discrimination using principal component analysis. Journal of Electroanalytical Chemistry, 593(1):105– 110, 2006.

[35] Katherine A Mirica, Joseph M Azzarelli, Jonathan G Weis, Jan M Schnorr, and Timothy M Swager. Rapid prototyping of carbon-based chemiresistive gas sensors on paper. Proceedings of the National Academy of Sciences, 110(35):E3265–E3270, 2013.

[36] Pavel V Nikitin and KVS Rao. Antennas and propagation in UHF RFID systems. Challenge, 22:23, 2008.

[37] JP Novak, ES Snow, EJ Houser, D Park, JL Stepnowski, and RA McGill. Nerve agent detection using networks of single-walled carbon nanotubes. Applied physics letters, 83(19):4026–4028, 2003.

[38] Cecilia Occhiuzzi, Amin Rida, Gaetano Marrocco, and Manos Tentzeris. Rfid passive gas sensor integrating carbon nanotubes. Microwave Theory and Techniques, IEEE Transactions on, 59(10):2674–2684, 2011.

[39] SV Patel, TE Mlsna, B Fruhberger, E Klaassen, S Cemalovic, and DR Baselt. Chemicapacitive microsensors for volatile organic compound detection. Sensors and Actuators B: Chemical, 96(3):541–553, 2003.

[40] Pengfei Qi, Ophir Vermesh, Mihai Grecu, Ali Javey, Qian Wang, Hongjie Dai, Shu Peng, and KJ Cho. Toward large arrays of multiplex functionalized carbon nanotube sensors for highly sensitive and selective molecular detection. Nano letters, 3(3):347–351, 2003.

[41] Angel Ramos, David Girbau, Antonio Lazaro, Ana Collado, and Apostolos Georgiadis. Solar-powered wireless temperature sensor based on UWB RFID with self-calibration. Sensors Journal, IEEE, 15(7):3764–3772, 2015.

[42] Alanson P Sample, Jeff Braun, Aaron Parks, and Joshua R Smith. Photovoltaic enhanced UHF RFID tag antennas for dual purpose energy harvesting. In RFID (RFID), 2011 IEEE International Conference on, pages 146–153. IEEE, 2011.

[43] Alanson P Sample, Daniel J Yeager, Pauline S Powledge, and Joshua R Smith. Design of a passively-powered, programmable sensing platform for UHF RFID systems. In RFID, 2007. IEEE International Conference on, pages 149–156. IEEE, 2007.

[44] Sudhir Shrestha, Mercyma Balachandran, Mangilal Agarwal, Vir V Phoha, and Kody Varahramyan. A chipless RFID sensor system for cyber centric monitoring applications. Microwave Theory and Techniques, IEEE Transactions on, 57(5):1303–1309, 2009.

[45] Karl Skucha, Zhiyong Fan, Kanghoon Jeon, Ali Javey, and Bernhard Boser. Palladium/silicon nanowire schottky barrier-based hydrogen sensors. Sensors and Actuators B: Chemical, 145(1):232–238, 2010.

[46] ES Snow, FK Perkins, EJ Houser, SC Badescu, and TL Reinecke. Chemical detection with a single-walled carbon nanotube capacitor. Science, 307(5717):1942–1945, 2005.

[47] Alexander Star, Vikram Joshi, Sergei Skarupo, David Thomas, and Jean-Christophe P Gabriel. Gas sensor array based on metal-decorated carbon nanotubes. The Journal of Physical Chemistry B, 110(42):21014–21020, 2006.

[48] Joseph R Stetter and Jing Li. Amperometric gas sensors a review. Chemical reviews, 108(2):352–366, 2008.

[49] Joseph R Stetter, Edward F Stetter, Daniel Ebeling, Melvin Findlay, and Vinay Patel. Printed gas sensor, November 24 2010. US Patent App. 12/953,672.

[50] Vivek Subramanian, Jean MJ Fr'echet, Paul C Chang, Daniel C Huang, Josephine B Lee, Steven E Molesa, Amanda R Murphy, David R Redinger, and Steven K Volkman. Progress toward development of all-printed RFID tags: materials, processes, and devices. Proceedings of the IEEE, 93 (7):1330–1338, 2005.

[51] Arnaud Vena, Lauri Sydanheimo, Manos M Tentzeris, and Leena Ukkonen. A novel inkjet printed carbon nanotube-based chipless RFID sensor for gas detection. In Microwave Conference (EuMC), 2013 European, pages 9–12. IEEE, 2013.

[52] Juha Virtanen, Leena Ukkonen, Toni Bj"orninen, Atef Z Elsherbeni, and Lauri Sydanheimo. Inkjet-printed humidity sensor for passive UHF RFID systems. Instrumentation and Measurement, IEEE Transactions on, 60(8):2768–2777, 2011.

[53] J.A. Vitaz, A.M. Buerkle, M. Sallin, and K. Sarabandi. Enhanced detection of on-metal retro-reflective tags in cluttered environments using a polarimetric technique. Antennas and Propagation, IEEE Transactions on, 60(8):3727–3735, Aug 2012.

[54] Rushi Vyas, Vasileios Lakafosis, Amin Rida, Napol Chaisilwattana, Scott Travis, Jonathan Pan, and Manos M Tentzeris. Paper-based RFID-enabled wireless platforms for sensing applications. Microwave Theory and Techniques, IEEE Transactions on, 57(5):1370–1382, 2009.

[55] Kirsi Wallgren and Sotiris Sotiropoulos. Oxygen sensors based on a new design concept for amperometric solid state devices. Sensors and Actuators B: Chemical, 60(2):174–183, 1999.

[56] Joshua Ray Windmiller and Joseph Wang. Wearable electrochemical sensors and biosensors: a review. Electroanalysis, 25(1):29–46, 2013.

[57] Noboru Yamazoe. New approaches for improving semiconductor gas sensors. Sensors and Actuators B: Chemical, 5(1-4):7–19, 1991.

[58] Li Yang, Rongwei Zhang, Daniela Staiculescu, CP Wong, and Manos M Tentzeris. A novel conformal RFID-enabled module utilizing inkjet- printed antennas and carbon nanotubes for gas-detection applications. Antennas and Wireless Propagation Letters, IEEE, 8:653–656, 2009.

[59] Hyeun Joong Yoon, Jin Ho Yang, Zhixian Zhou, Sang Sik Yang,

Mark Ming-Cheng Cheng, et al. Carbon dioxide gas sensor using a graphene sheet. Sensors and Actuators B: Chemical, 157(1):310–313, 2011.

[60] Jie Zhao, Yinghua Yu, Bo Weng, Weimin Zhang, Andrew T Harris, Andrew I Minett, Zhilian Yue, Xu-Feng Huang, and Jun Chen. Sensitive and selective dopamine determination in human serum with inkjet   printed   nafion/MWCNT   chips.   Electrochemistry Communications, 37:32–35, 2013.

# CHAPTER TWO

## Automated Robust Design Optimization of Analog Circuits for Flexible Electronics

Chien-Nan Jimmy Liu, Yen-Lung Chen, and Nguyen Cao Qui

*Department of Electrical Engineering, National Central University, Taiwan (ROC)*

## Abstract

Circuit design for flexible electronics is still a challenging task, especially for sensitive analog circuits. Due to the different properties of flexible thin-film transistors (TFTs), conventional CMOS design techniques cannot be used directly in flexible electronics. Significant parameter variations and degradation effects of flexible TFTs further increase the difficulties for circuit designers. In this chapter, an incremental Latin hypercube sampling (LHS) technique is presented to speed up circuit simulation while considering process variation and aging effects by reusing previous sampling information. And, a reliability-aware circuit sizing approach is proposed for the analog circuits with flexible TFTs. The process variation, bending, and degradation effects of flexible TFTs in the optimization flow are considered simultaneously. Instead of optimizing the fresh yield and lifetime yield separately, this work proposes a unified optimization approach to consider the two yield issues simultaneously. As shown in the experimental results, the proposed approach can further improve the lifetime yield and significantly reduce the design overhead with a fast computation time.

**Keywords**: Flexible electronics, Process variation, Aging effect, Latin hypercube sampling.

## 1. Introduction

Flexible electronics are a possible alternative to conventional silicon electronics for portable consumer applications with many advantages. The most commonly used active device in flexible electronics is the thin-film transistor (TFT). However, the circuit design for flexible electronics is challenging for several reasons [1]. Due to quite different properties of flexible TFTs, conventional CMOS design techniques cannot be used directly in flexible electronics. The most challenging issue for designers is the unstable mobility of TFTs due to their flexible property. Many research studies have shown that the mechanical strain may change the TFT's mobility significantly. In conventional CMOS technology, the transistor mobility is almost fixed. Therefore, a robust circuit design methodology is necessary for flexible electronics.

Besides those fresh yield issues, the degradation effects are more significant in flexible TFT processes than those in silicon CMOS processes. The behaviors of flexible TFTs may change a lot with degradation, which make the design fail to meet its specifications in just a few seconds. This is often called the reliability issue or the lifetime issue. Therefore, a robust circuit design methodology that considers both the fresh yield and the time-dependent reliability is essential to implement more complex applications of flexible electronics.

To understand the uncertainty due to the parameter variations, statistical circuit analysis is required to analyze the circuit during the pre-silicon phase. It helps designers improve their design yield at an early stage and reduce re-design cycles and re-spin cost. As the golden standard method of statistical circuit analysis, Monte Carlo (MC) method draws massive samples according to the process variations and output the statistical distribution of the circuit performance. However, with aging effects, it is challenging to calculate the lifetime of a circuit while considering process variation and performance degradation simultaneously. The total runtime can be prohibitively long when MC simulation is required at each aging time step.

Unlike MC simulation that uses a massive random sample; several existing sampling methods have been proposed to achieve acceptable coverage with fewer samples. Instead of using random samples, Quasi-Monte Carlo (QMC) [2], and Latin Hypercube Sampling (LHS) [3] methods use a low-discrepancy sequence (a.k.a. quasi-random sequence) to obtain faster coverage rate in most cases. Although the sampling methods alleviate the simulation burden in MC analysis a little bit, repeated simulations are still required at each time step due to the transistor aging over time. Therefore, an incremental LHS technique [4] is proposed to reduce the number of samples that are required to be simulated when the time step updates. By reusing the information of previous samples, only a small portion of samples are incrementally updated, which significantly improve the efficiency of aging analysis.

After reliability analysis, some design-for-yield (DFY) techniques are required to consider the parameter variations in the design phase, especially for analog circuits. By carefully adjusting the device sizes and nominal design point of the circuit, the tolerance to parameter variations can be improved significantly. Because DFY techniques keep the original circuit structure and simply adjust the device sizes, the incurred circuit overhead could be smaller than extra compensation circuitry. For flexible TFTs, the degradation effects are much more significant than those in silicon CMOS. Simple linear prediction may introduce large errors while estimating the performance with degradation in flexible electronics. Therefore, those DFY techniques for CMOS circuits cannot be applied to flexible electronics directly. In this work, a unified robust optimization flow is proposed to consider the parameter variation and aging issues simultaneously for the different properties of flexible TFTs. As shown in the experimental results, both fresh yield and lifetime yield can be significantly improved with less over-design.

## 2. Design Issues with Flexible TFTs

### 2.1 Mechanical Stress on Flexible TFTs

Due to their flexibility, flexible TFTs have more serious impacts from the mechanical stress than those in conventional CMOS. The strain force significantly changes the electron mobility and the current of each transistor as time goes by. As illustrated in the left of Fig. 1 [5], the compressive stress force on the flexible TFT squeezes this transistor. Because the effective channel length of this transistor is decreased under the compressive mode, its mobility will be increased. As illustrated on the right half of Fig. 1, the tensile stress force on the flexible TFT stretches this transistor, which increases its effective channel length and decreases its mobility. Fig. 2 [6] shows the measured mobility change of pMOS and nMOS under longitudinal stress and transversal stress. The mobility is changed more significantly under the transversal stress (tensile mode) than in the longitudinal stress (compressive mode). The mobility change of pMOS is different from the change of nMOS. Under the tensile strain (outward bent), the maximum mobility change is + 7.5%. Under the compressive strain (inward bent), the maximum mobility change is −25%. This result shows that the mobility change of flexible TFTs is not a symmetrical distribution under mechanical stress, which violates the assumption of normal distribution in previous works. Therefore, the log-normal distribution is adopted in this work to model the bending effects. All the assumptions and probability calculations are redefined in this work to deal with nonsymmetrical variations.

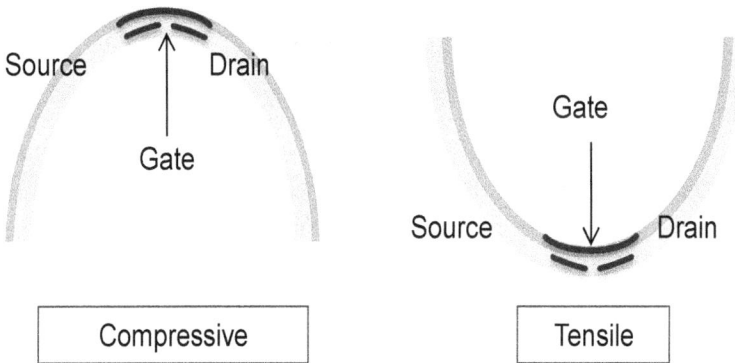

Figure 1: Different bending schemes in flexible TFTs.

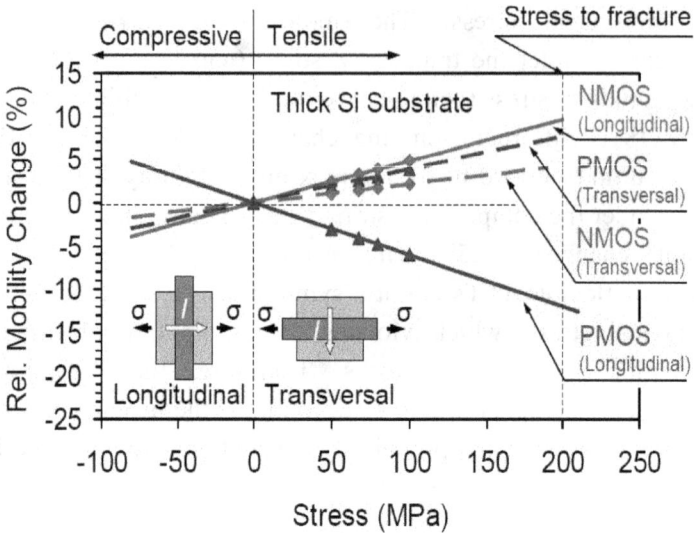

Figure 2: Mobility change of poly-Si TFTs under stress.

## 2.2 Transistor Aging of Flexible TFTs

Besides the bending effects, the parameter variation due to transistor aging is also a big issue of flexible TFTs. The primary reason responsible for the threshold voltage degradation of flexible TFTs is the charge injection in the silicon nitride (SiNx) gate insulator. Given a positive bias to the gate electrode, the energy bands are changed to allow the electrons traveling from source to drain via channels. However, the hot carriers will be injected into the gate gradually, which decreases the source-drain current. Due to the charge redistribution, the threshold voltage becomes a time-dependent variable, not a fixed value as previously assumed. As shown in Fig. 3 [6], the measured threshold voltage shift of a-Si TFTs  is quite large (>20%), which can no longer be ignored.

Compared with conventional CMOS technology, the aging effect is more critical in flexible TFTs. In conventional CMOS, the degradation effects have to be observed after a few years. However, in flexible TFTs, the same degradation can be reached in just a few seconds, as shown in Fig. 3. If this degradation is not properly considered, the design may fail to meet its specifications after a short time. Therefore, it is important to consider both the fresh yield and the time-dependent reliability when designing the circuits with flexible TFTs.

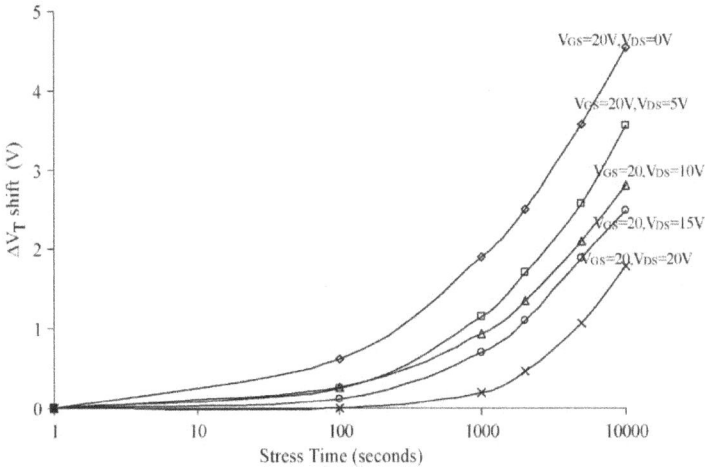

Figure 3: Threshold voltage degradation of a-Si TFTs.

The transistor parameter that has the most significant degradation is the threshold voltage ($V_{th}$). In order to simplify the degradation analysis, most of the previous works assume that only $V_{th}$ is changed over time. The same assumption is also adopted in this work. The change of the threshold voltage of a-Si thin-film transistors (TFT) ($\Delta V_{th}^{t_s}$) with degradation at stress time ($t_s$) can be expressed as Eq. (1) and (2) [7]. In these equations, the stress time $t_s$ is the time period after the initial time ($t_0$), which is a user given value that represents the target lifetime. $\alpha$ and $\beta$ are two process-dependent parameters. According to the reported data in [7], this degradation model shows good consistency with real measurement results.

$$\Delta V_{th}^{t_s} = (V_{gs} - V_{th}) \cdot \{1 - \exp[-\left(\tfrac{t_s}{\tau}\right)^{\beta}]\} \qquad (1)$$

$$\tau = k \cdot (V_{gs} - V_{th})^{\frac{1-\alpha}{\beta}} \qquad (2)$$

## 3. Increment Sampling for Variation and Lifetime Analysis

In fact, circuit aging is a gradual process, which does not change the transistor behavior drastically in each time step. Hence, there could be a large portion of overlap between the sample space in a time step and the one after. Based on this observation, an incremental Latin hypercube sampling (LHS) approach is proposed [8] to reduce the number of samples that are required to be simulated when time step updates. Instead of updating and re-simulating all the samples, we only incrementally update a small portion of samples.

## 3.1 Analysis Flow

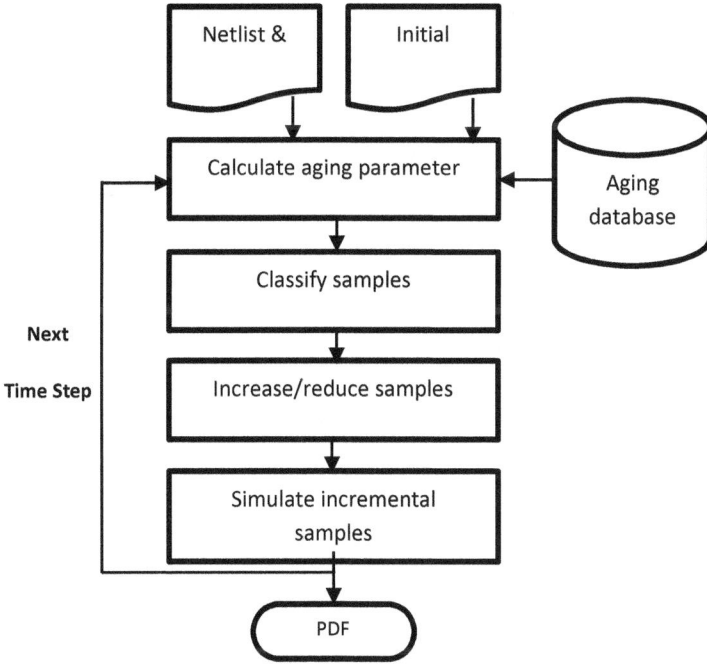

Figure 4: Flowchart of the incremental LHS for circuit analysis.

Considering the process variations, the circuit parameters, such as threshold voltage ($V_{th}$) and transistor width ($W$), are modeled as random variables according to the models provided by the foundry. Those parameters form a random parameter vector $S = [X_1, X_2, ..., X_N]$. The distribution of the circuit performance metric $Y$ is derived by simulating a large number of circuits (a.k.a. samples) whose parameter vectors are $\{s_1, s_2, ..., s_M\}$ respectively.

In addition, the aging effect impacts these circuit parameters and introduces another dimension to the analysis. To understand the life cycle of a design, designers need to analyze the model of a full aging history, which requires to simulate $M$ samples at every aging time step, and leads to prohibitively long simulation time.

The overall flow of the proposed algorithm is illustrated in Fig. 4. First, the circuit information such as parameter distribution, circuit performance and PDF will be initialized at the first step. In the next step, new parameters after aging will be calculated to obtain the performance degradation. Then, Latin Hypercube Samples are obtained to generate the performance distribution for the initial PDF, and the new PDF in the next time step will be fitted based on the this PDF. In the next time step, the parameter space is adjusted by adding or removing some samples. Using this method, the property of LHS can be kept, and only partial samples need to be re-simulated.

## 3.2 Incremental Latin Hypercube Sampling

For each parameter, LHS divides the sample space into $M$ equally probable intervals and makes sure samples are evenly distributed in these intervals. Considering the aging effect, the distribution of a parameter may change over time. However, the distributions over the adjacent time steps will not change drastically because aging affects transistor parameters gradually. As an example, the original distribution of $V_{th}$ on a fresh 8 μm flexible TFT and the distribution after 100s aging of the same transistor are illustrated as $h^0$ and $h^{100}$ in Fig. 5. After 100s aging, we can observe that the distribution is only slightly shifted to the right hand side, while there is a large portion of overlap between these two distributions.

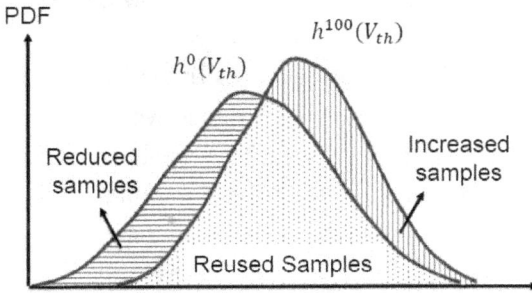

Figure 5: Overlap of the original $V_{th}$ distribution and that after aging.

Based on the observation, we can obtain the new distribution in time=100s by adding a few new samples to those original samples and removing some redundant samples from them, while keeps the majority of samples. That is the motivation of the proposed incremental LHS sampling. More importantly, it is easy to keep the property of LHS in the incremental sampling, i.e. these samples are evenly distributed in each equally probable intervals in each dimension. This ensures the coverage in the sample space. In the following discussion, we denote the $V_{th}$ as the $i^{th}$ parameter, $X_i$, in the parameter vector $S$ without losing generality.

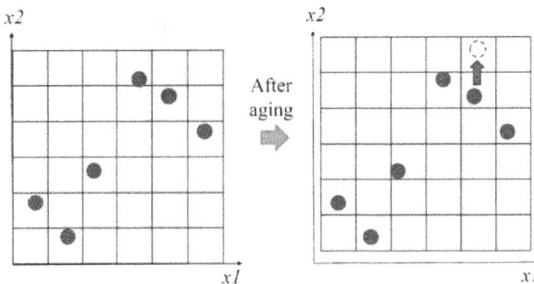

Figure 6: Incremental Latin hypercube sampling with 2 dimensions.

Take Fig. 6 as an example. Same to those $M$ initial LHS samples before aging, we divide $M$ equally probable intervals after one aging time step. Ideally, there is exactly one sample per interval. Because of the distribution shifting, however, an interval may cover multiple samples or no sample. In this case, the incremental LHS removes the "extra" samples in some intervals and adds them to empty intervals by incremental sampling, i.e. changing the $X_i$ value of the "extra" sample into a random value in the empty interval.

### 3.3  Selecting Incremental Samples

To determine the samples to be increased, reused and reduced, we use a greedy approach in this work. The pseudocode is elaborated in **Algorithm 1**. Assume there are $N$ parameters $S = [X_1, X_2, ..., X_N]$ with a total of $M$ samples $\{s_1, s_2, ..., s_M\}$, where the $i^{th}$ sample $s_i$ is an $N$-dimension vector $[x_{i1}, x_{i2}, ..., x_{iN}]$. **Algorithm 1** iterates all the $N$ dimensions and makes sure those $M$ samples are distributed in $M$ intervals with equal probability. In each dimension, the algorithm first locates the "extra" samples, and relocates them to the interval that does not have any samples in line 16 of **Algorithm 1**. Since it only changes the value $x_{ji}$ in parameter $i$, it does not change the distribution on other parameters. Therefore, by iterating this process for each parameter, the new samples still satisfy the property of LHS by any means.

---

**Algorithm 1** Sample selection algorithm for incremental for LHS

**Input:**　　1) Old samples $s_1, s_2, ..., s_M$
　　　　　　2) New distribution of each dimension $X_1, X_2, ..., X_N$
**Output:**　1) A set of incremental samples $S_{inc} = \{s_{11}, s_{12}, ...\}$
　　　　　　2) A set of indices of the samples to be reduced $I_{reduce}$

**Algorithm:**
1: $I_{reduce} = \{\}$
2: **for** $\forall i \in \{1, 2, ..., N\}$ **do**
3:　　Divide $X_i$ into $M$ intervals with equal probability
4:　　$Cnt_k = 0, \forall k \in \{1, 2, ..., M\}$
　　　// Find out the "extra" samples
5:　　$I_{local} = \{\}$
6:　　**for** $\forall j \in \{1, 2, ..., M\}$ **do**
7:　　　　Check sample $s_j$ belongs to interval $k$
8:　　　　**if** $Cnt_k = 1$ **then**
9:　　　　　　Add $j$ to set $I_{local}$
10:　　　**else**
11:　　　　　$Cnt_k = 1$
12:　　　**end if**
13:　　**end for**
　　　// Incremental sampling
14:　　**for** $\forall j \in I_{local}$ **do**
15:　　　Find the first $k$ with $Cnt_k = 0$
16:　　　Set $x_{ji}$ to a random value in interval $k$
17:　　　$Cnt_k = 1$
18:　　**end for**
19:　　Add $I_{local}$ to $I_{reduce}$
20: **end for**
　　　// Collect all the incremented samples
21: Add $s_j$ to $S_{inc} \forall j \in I_{reduce}$

---

Figure 7: Algorithm 1: Sample selection for incremental LHS.


## 3.4 Experimental Results

In order to verify the incremental sampling technique, an OPA circuit [9] shown in Fig. 8 implemented with ITRI a-Si 8μm process, is used to perform some comparisons. The relative standard deviation (*std*) of $V_{th}$ and μ are modeled as ±20% and ±15% respectively, while the relative *std* of device physical sizes (*W* and *L*) are modeled as ±3%. We analyze the aging effects with target lifetime of 1000, and 10000 seconds, while the aging time step is configured as 100 seconds. The results are shown in Table 1.

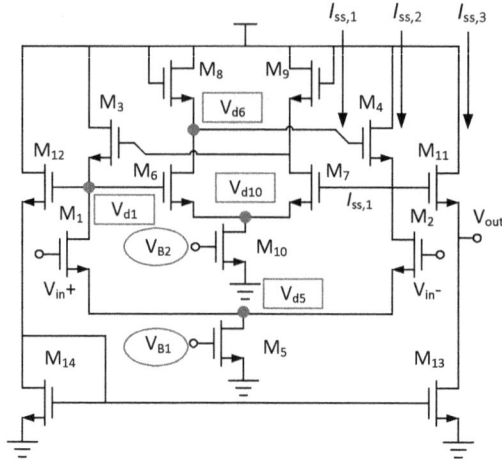

Figure 8: The schematic of the operational amplifier with a-Si TFTs.

Table 1: Comparison of Sampling Approaches

| Time (sec.) | | MC(6K) | Quad | iLHS |
|---|---|---|---|---|
| 0 | Accuracy | 100% | 99% | 99% |
| | #Samples | 6000 | 2000+2000 | 500 |
| 1000 | Accuracy | 100% | 85% | 99% |
| | #Samples | 6000 | - | 16 |
| 10000 | Accuracy | 100% | 74% | 99% |
| | #Samples | 6000 | - | 29 |

| Overall | #Samples | 600000 | 4000 | 2469 |
|---|---|---|---|---|
| | Speedup | 1X | 150X | 243X |

All the circuits are simulated using HSPICE. Instead of measuring the actual simulation time, we count the number of simulation runs to calculate the speedup. We also implement the following approaches for comparison purpose:

a) MC method: Calculate the probabilistic distributions from a large number of random samples. At each time step, all these samples need to be re-simulated with the updated transistor parameters. In particular, we implement two MC simulations, one with 6000 samples to serve as the golden standard.

b) Quad Model for WCD [9]: Quadratic degradation model is used to predict the parameter variations with aging effects.

c) iLHS (the proposed method): Reliability and variability are considered simultaneously. The incremental LHS is used to reduce the re-simulation cost efficiently.

In incremental LHS method, 500 samples are used to build the behavior model at the initial time step (*time*=0s). Benefit from the incremental sampling, only a few samples are generated and simulated at each time step. Compared to the MC results, 243x average speedup can be obtained from t=0s to t=10000s. The accuracy of PDF estimation can still be kept because the property of LHS can be guaranteed in the proposed approach.

The Quad model requires 2000 samples to build the performance model, and the other 2000 samples are required to obtain the aging ratio. However, the accuracy of the quadratic model is still lower than the proposed method because of the prediction error of the non-linear aging rate.

## 4.   Simultaneous Optimization with Reliability and Variability

In the literature, many robust optimization techniques for CMOS analog circuits are proposed based on simulation-based synthesis approaches [12]. Although these approaches can achieve quite accurate results, they often have complexity due to the long simulation time in the optimization process. As reported in previous studies, several hours are often required to optimize a typical-sized circuit. Equation-based optimization, such as geometric programming (GP)-based analog circuit synthesis [13], is another popular approach with fast computation time. In [14] and [15], the traditional design centering concept is included in the GP-based analog circuit synthesis by maximizing the distance to all the boundaries. However, as studied in the related research [16], the center of the feasible region is not necessarily the nominal point with maximum yield. Because the process sensitivity of different circuit performances can be different, the weight of the distance to each feasible boundary should also be different.

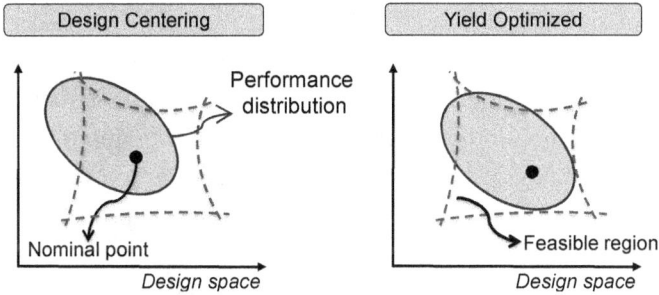

Figure 9: Yield optimization with different process sensitivity

In [16] and [17], the concept of worst-case distance (WCD) is introduced to provide a quick estimation of the design yield. A similar concept is also applied to the performance space, which is called the process capability (Cpk) in [12], for yield evaluation. The key idea is normalizing the distance to the feasible boundary by the standard deviation ($\sigma$) of its performance distribution to consider different process sensitivity. Using this metric to evaluate the design yield may be more accurate than using the row distance only. As illustrated in Fig. 9, finding a proper nominal point rather than the center may push more performance distribution into the feasible region to gain higher yield. Nevertheless, the performance distribution and process sensitivity may vary at different nominal points. The corresponding Cpk should also be recalculated at different nominal points, which implies a reevaluation of the statistical distribution at every possible solution. Since Cpk cannot be calculated as a fixed equation, it can only be used with the simulation-based optimization to guide the yield-aware analog circuit synthesis, which often requires long computation time.

If the time-dependent reliability issue is also considered in the yield-aware synthesis flow, it is similar to considering the design yield repeatedly at many time frames because the degradation is an accumulated behavior. This significantly increases the complexity of the problem. In a previous report [18], several hours are often required to analyze the degradation effects upon circuit behaviors. In the literature, many speed-up approaches are proposed for degradation analysis. However, the cost is still too expensive to adopt the simulation-based approaches for considering the fresh yield and lifetime yield simultaneously.

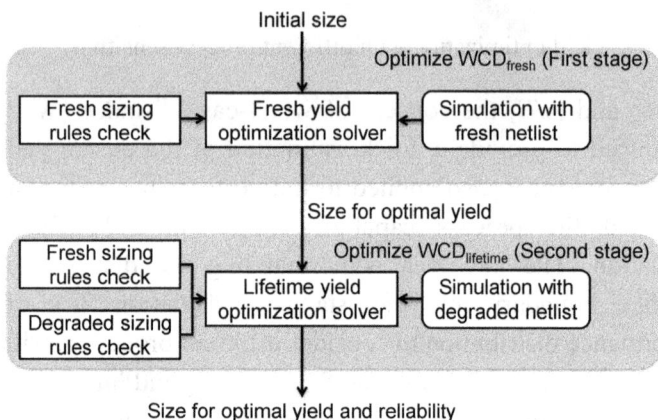

Figure 10: Yield optimization with different process sensitivity.

For silicon CMOS circuits, the DFY technique with reliability consideration is proposed [19]. This method is a two-stage approach, as illustrated in Fig. 10. The first stage considers only the fresh yield without degradation to offer a better initial solution for the second stage. In other words, the objective of this stage is maximizing the $WCD_{fresh}$ value. In the second stage, a linear model is used to predict the wors tcase distance of the design with degradation. The circuit optimization is then performed again based on the predicted WCD values, which is called the lifetime yield optimization in. A quadratic aging model is then proposed [9] to improve the WCD prediction accuracy. Without simulating the performance with degradation, this approach significantly reduces the computation cost of performance calculation and aging analysis. As reported in the paper [9], the reliability-aware optimization can be finished in minutes by using this approach. However, after lifetime yield optimization, the optimal point of stage 1 will be changed. The fresh yield may no longer be the best value.

To solve these issues, a unified robust optimization flow is proposed for the analog circuits with flexible TFTs [11]. The different properties of flexible TFTs, such as the large $V_{th}$ variation, bending and aging effects, are considered in the simultaneous optimization algorithm. Because the mobility change of the flexible TFTs under bending is not a symmetrical distribution and violates the assumption of normal distribution in previous works, a new yield evaluation term, cumulative success probability (CSP), is defined with the corresponding probability calculations to deal with non-symmetrical variations. As shown in the experimental results, accurate prediction of performance distribution can lead to a point with optimized results and less over-design.

## 4.1 Variation-aware Optimization Flow

Fig. 11 shows the proposed variation-aware optimization flow for the analog circuit sizing problem in flexible electronics. Basically, it is built on the bias-driven optimization flow for analog circuits [21]. Given the circuit database (netlist, performance equations, and process information) and the required specifications, the constraints and cost function for the mathematical solver are set up at the first step. Other design constraints, such as symmetrical or proportional constraints, are also added to this step. After that, the optimized bias voltages and bias currents that make the circuit satisfy all specifications with optimized cost are identified by a linear programming (LP) solver with the prebuilt tables of transistor parameters. The yield of each possible design point is also calculated directly in the mathematical solver to optimize the design yield simultaneously. Finally, the size of each transistor is determined by the bias voltage and bias current from a prebuilt sizing table. Compared to the original optimization flow in [21], the major difference is the equation-based variance analysis in the optimization engine. This step will be explained as follows.

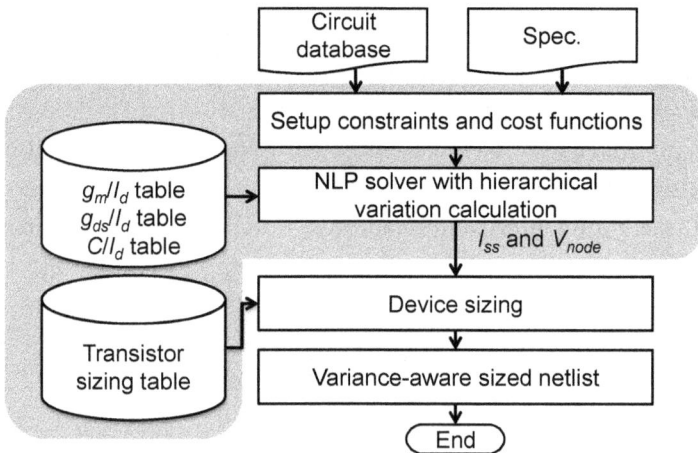

Figure 11: Variation-aware Optimization Flow.

Because the different process sensitivities are not considered in previous approaches [16], maximizing the row distances to all feasible boundaries may not achieve the highest yield. In order to normalize the distance to each feasible boundary by its standard deviation ($\sigma$), the performance distribution at present nominal points should be evaluated first. The typical approach is to collect the statistical data through simulating many instances of the circuit. However, this approach requires large computation efforts and cannot be used with equation-based optimization to provide fast-yield optimization.

In this part, an equation-based variation analysis method is proposed to calculate the variance of each performance in the circuits without simulations. Although there are some analytical variance calculation methods [22] proposed in the literature, they all assume that the probability distribution is symmetrical normal distribution. In this section, a new yield evaluation term, cumulative success probability (CSP), is defined in (3) based on the cumulative distribution function (CDF) to deal with non-symmetrical distribution. The conception of CSP is illustrated in Fig. 12, and the evaluation term. The contrary of CSP is called cumulative failure probability (CFP) as defined in (4). In order to obtain better design yield, the CSP should be maximized, while the CFP should be minimized. All the calculation steps are modified to deal with CDF because CDF is also applicable to symmetrical normal distribution, this approach greatly improves the generality on different kinds of parameter variations.

$$CSP = 1 + \frac{1}{\sigma\sqrt{2\pi}} \int_{-\infty}^{spec} exp^{\frac{(t-N)^2}{2\sigma^2}} dt \qquad (3)$$

$$CFP = 1 - CSP \qquad (4)$$

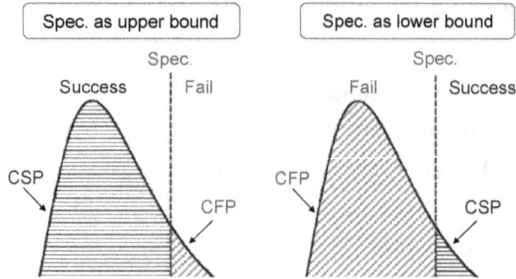

Figure 12: Illustration of CSP and CFP.

In order to calculate the CSP, the system behavior is modeled as a 3-level hierarchy without losing generality, as shown in Fig. 13. The bottom level represents the variations of process parameters, such as width (W), length (L), threshold voltage ($V_{th}$), mobility ($\mu$), etc. Those data can be obtained from the foundry-given process model. The intermediate level represents the variations of transistor parameters, such as transconductance ($g_m$), drain current ($I_D$), output conductance ($g_{ds}$), etc. Those variations are calculated from the process-level variations. The top level represents the variations of system-level performances, such as gain, bandwidth, phase margin, etc. Those variations are calculated from the intermediate-level variations.

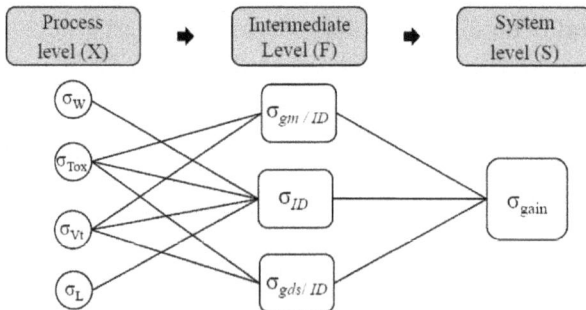

Figure 13: Hierarchical variance calculation.

Because the process-level variations are given as independent random variables in the process model, they are also assumed to be independent variables. For each intermediate parameter $(F_i,$ $C_{X_i}^{F_j})$ is defined as its partial derivative with respect to a process parameter $X_i$, as shown in Eq. (5). By using the first order Taylor series expansions, the intermediate parameter $F_i$ can be written as Eq. (6). The variance of this intermediate parameter $(\sigma^2 F_j)$ can be defined as Eq. (7), in which $X_i$ is the mean value of the process parameter $X_i$. Once the present nominal point of the design is determined, the partial derivation and the mean value of each $X_i$ are determined. It implies that the variance of each intermediate parameter can be calculated through the following equations without simulations.

$$C_{X_i}^{F_j} = \frac{\partial F_j}{\partial FX_i}\bigg\| (X_1, X_2, \dots) \tag{5}$$

$$F_j = F_j(\mu_{X1}, \mu_{X2}, \dots) \tag{6}$$

$$\sigma_{F_j}^2 = E\left\{\left|\sum_{i=1}^{n_X} C_{X_i}^{F_j}(X_i - \mu X_i)\right|^2\right\} = \sum_{i=1}^{n_F} C_{X_i}^{F_i} C_{X_i}^{F_i} \sigma_{X_i}^2 \tag{7}$$

$$\sigma_H^2 = \sum_{j=1}^{n_F} \sigma_{F_j}^2 + CCT \tag{8}$$

$$CCT = \sum_{i=1}^{n_X} \sum_{j,h=1}^{n_F} \left(C_{F_i}^H C_{X_i}^{F_j}\right)\left(C_{F_h}^H C_{X_i}^{F_h}\right) \tag{9}$$

Unlike the process parameters, the intermediate-level parameters are often highly correlated to each other. While calculating the variations of system-level performances from the intermediate-level variations, the correlations between those transistor parameters should be properly considered. Therefore, an additional correlation correction term (CCT) is required in the variance calculation of system-level performances, as shown in Eq. (8). This CCT term can be derived from the intermediate-level parameters that are related to this system-level parameter $S_i$, as shown in Eq. (9). The overall CSP calculation algorithm is shown in Fig. 14.

---

**Algorithm 2: Hierarchical CSP Calculation**

**Input :** Feasible Design Space ($D_f$), technology files, circuits database, nominal point ($N$)

**Output :** CSP of each performance

---

| | |
|---|---|
| 1 | **for** $i$=1 to $n_x$ **do** // for each process parameter |
| 2 | $\sigma^2_{X_i} \leftarrow$ technology files |
| 3 | **end for** |
| 5 | // calculate the variance ($\sigma^2_{F_j}$) of intermediate parameter ($F_j$) |
| 6 | **for** $j$=1 to $n_f$ **do** // for each intermediate parameter |
| 7 | **for** $i$=1 to $n_x$ **do** // for each process parameter |
| 8 | $V_{i,j} \leftarrow C_{X_i}^{F_j} \cdot C_{X_i}^{F_j} \cdot \sigma^2_{X_i}$ |
| 9 | **end for** |
| 10 | $\sigma^2_{F_j} \leftarrow \sum_{i=1}^{n_i}(V_{i,j})$ |
| 11 | **end for** |
| 12 | **for** $k$=1 to $n_s$ **do** // for each system performance |
| 13 | // calculate the independent term |
| 14 | **for** $j$=1 to $n_f$ **do** // for each intermediate parameter |

| | |
|---|---|
| 15 | $V_{j,k} \leftarrow C_{F_j}^{S_k} \cdot \sigma_j^2$ |
| 16 | **end for** |
| 17 | $I_k \leftarrow \sum_{j=1}^{n_j}(V_{j,k})$ |
| 18 | // calculate the correlation correction term |
| 19 | **for** $h$=1 to n$_f$  **do** // for each intermediate parameter |
| 20 | **for** $j$=1 to n$_f$  **do** // for each intermediate parameter |
| 21 | **if** $h \neq j$ **then** |
| 22 | **for** $i$=1 to n$_x$ **do** // for each process parameter |
| 23 | $V_{i,j,h,k} \leftarrow (C_{F_j}^{S_k} \cdot C_{X_i}^{F_j}) \cdot (C_{F_h}^{S_k} \cdot C_{X_i}^{F_h}) \cdot \sigma_{X_i}^2$ |
| 24 | **end for** |
| 25 | **end if** |
| 26 | **end for** |
| 27 | **end for** |
| 28 | $CCT_k \leftarrow \sum_{i=1}^{n_i}\sum_{j,h=1,j \neq h}^{n_f}(V_{i,j,h,k})$ |
| 29 | $\sigma_{S_k}^2 \leftarrow I_k + CCT_k$ |
| 30 | $CFP = 1 - CSP(\sigma_{S_k}^2, N, Spec)$ |
| 31 | **end for** |
| 32 | **return** CSP of each performance |

Figure 14. The hierarchical variance analysis.

## 4.2  Variation-aware Optimization With Aging Model

In this work, the adopted $V_{th}$ aging model is an equation-based model as shown in Eq. (1-2), which can be obtained by either real measurement or repeatedly simulations by using the methods in Sec. II. In order to obtain the characteristic of each transistor and the circuit performance, the most straightforward method to is conducting a circuit simulation with the updated threshold voltage at $t_s$, which is defined as $V_{th}^{t_s}$ in Eq. (10). However, the circuit sizing flow in this part is an equation-based approach with pre-built lookup tables. Adding a re-simulation step at each time step is not efficient in this flow. If those lookup tables for transistor parameters are rebuilt for every different ts, the simulation cost is also too expensive to be implemented in the equation-based sizing flow.

$$V_{th}^{t_s} = V_{th}^{t_0} + \Delta V_{th}^{t_s} \tag{10}$$

In order to avoid the re-simulation cost, a time-dependent transistor model is proposed to obtain the transistor parameters at ts from the original lookup tables built at $t_0$. For example, the original table index to obtain the $g_m/I_D$ of a transistor at $t_0$ is $V_{GS}^{t_0}$, as shown in Eq. (11). When the threshold voltage of a transistor is changed at ts $V_{th}^{t_s}$ due to degradation, the table index can be converted to $V_{GS}^{t_0} - \delta V_{th}^{t_s}$, as shown in Eq. (12). It implies that the original tables based on $V_{th}^{t_0}$ are still applicable with the updated index. Once the $g_m/I_D$ value at ts is obtained from the table, solving the circuit equations again obtains the node voltages and the corresponding $V_{GS}^{t_s}$. Due to the changes of $g_m/I_D$, the current ($I_D$) of the transistor is also changed from $t_0$ to ts. Therefore, the current at $I_D^{t_s}$ with degradation is modified by the ratio of the $I_D$ equation as shown in Eq. (13), in which  is a process-dependent parameter. By using this index transformation, the lookup tables for transistor parameters can be reused to avoid extra simulation cost for considering degradation effects. The efficiency of the circuit sizing engine can still be kept while adding time-dependent transistor models.

$$\frac{g_m^{t_0}}{I_D^{t_0}} = \frac{2}{V_{GS}^{t_0} - V_{th}^{t_0}} \tag{11}$$

$$\frac{g_m^{ts}}{I_D^{ts}} = \frac{2}{V_{GS}^{ts}-V_{th}^{ts}} = \frac{2}{V_{GS}^{ts}-(V_{th}^{to}+\Delta V_{th}^{ts})} = \frac{2}{(V_{GS}^{ts}-\Delta V_{th}^{ts})-V_{th}^{to}} \qquad (12)$$

$$I_D^{ts} = I_D^{to} \cdot \frac{(V_{GS}^{ts}-V_{th}^{ts})^\gamma}{(V_{GS}^{to}-V_{th}^{to})^\gamma} \qquad (13)$$

With the degraded transistor parameters, the variance of each performance after aging is recalculated. Using this method, fresh CSP and aging CSP can be obtained fast and accurately, as shown in Fig. 15. The whole flow is finished within a short operation time as demonstrated in the experimental results.

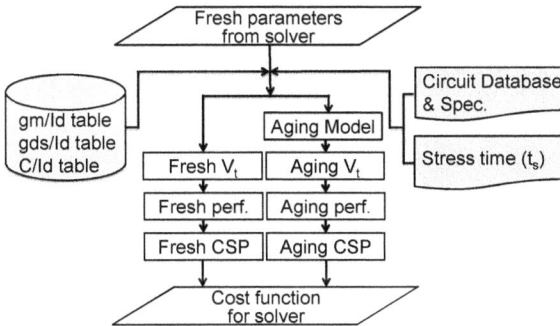

Figure 15. Hierarchical variance calculation flow with aging effects.

## 4.3 Experimental Results

In this section, we experiment on two circuits. The first case is the same OPA circuit in Sec. II.4 with ITRI a-Si 8-$\mu$m technology. The second case is a 4-bit digital-to-analog convertor (DAC) with the same a-Si process. Because flexible TFTs often require high supply voltage to provide enough electron mobility, the Vdd is set as 25V in the experiments. The circuit simulation results are obtained from HSPICE. In flexible electronics, the mobility ($\mu$) and threshold voltage ($V_{th}$) of TFTs can have large variations. Referring to the reported data in previous works, the variations of these two parameters are assumed to be −25% to + 10% (for $\mu$) and ±20% (for $V_{th}$) respectively. The variations of the device sizes (W and L) are assumed as ±3%.

## A) Accuracy Verification

The first experiment verifies the accuracy of the proposed performance model and variation analysis in the OPA case. In Table 2, Cases 1 and 2 are two synthesized OPAs with two different specifications. In each column of Table 2, M and $\sigma$ are the nominal points and its variance of each performance distribution with the assumed process variation model. Also, the $CDF_{95}$ is the target performance when the cumulative distribution achieves 95%. The row "Prediction" shows the prediction results in the synthesis engine based on the proposed hierarchical variance analysis model. The row "Simulation" shows the results of 1000-runs of Monte Carlo analysis using HSPICE with the synthesized netlist. The last row "Error" shows the error percentage between the prediction results and real simulation results. The computation time for the variance analysis is also reported for readers' reference.

According to the results of four key performances, gain, gain-bandwidth product (GBW), slew rate (SR) and phase margin (PM), the prediction of circuit performance using the proposed method is very close to the HSPICE simulation results as shown in Table 2. The error of each nominal performance is less than 5% compared to the HSPICE results. The error of the predicted variance is less than 7% compared to the HSPICE Monte Carlo analysis results, which demonstrates the effectiveness of the proposed hierarchical variation analysis. With accurate prediction of performance distribution, the variation-aware synthesis can be guided to a point that is much closer to the real optimized result.

The second experiment verifies the accuracy of the degradation model on the same circuit in the first experiment. The degradations on the threshold voltages of six key transistors at 10s, 100s and 1000s are shown in Table 3. The error of the predicted threshold voltage of each transistor is within 1%. Fig. 16 shows the threshold voltage change of each transistor from 0 to 1000s. The dots are the simulation results gained by using the golden model [18], and the solid lines represent the adopted model [7]. These results have demonstrated that the performance with degradation can be predicted accurately by using the adopted model without simulation.

Table 2: Accuracy verification of performance distribution.

| Technology / Performance | | 8μm a-Si technology | |
|---|---|---|---|
| | | Case 1 | Case 2 |
| Gain (dB) | Prediction | M=15.13 $\sigma$=1.21 $CDF_{95}$=17.9 | M=23.52 $\sigma$=2.31 $CDF_{95}$=28.9 |
| | Simulation | M=15.21 $\sigma$=1.23 $CDF_{95}$=18.0 | M=22.96 $\sigma$=2.29 $CDF_{95}$=28.3 |

|  |  |  |  |
|---|---|---|---|
|  | Error (%) | M=-0.53<br>σ=-1.63<br>CDF$_{95}$=0.71 | M=2.44<br>σ=0.87<br>CDF$_{95}$=-2.10 |
| GBW<br>(kHz) | Prediction | M=2.52<br>σ=0.031<br>CDF$_{95}$=2.60 | M=3.01<br>σ=0.072<br>CDF$_{95}$=3.12 |
|  | Simulation | M=2.61<br>σ=0.033<br>CDF$_{95}$=2.68 | M=2.89<br>σ=0.077<br>CDF$_{95}$=3.10 |
|  | Error (%) | M=-3.45<br>σ=-6.06<br>CDF$_{95}$=3.60 | M=4.15<br>σ=-6.49<br>CDF$_{95}$=-3.41 |
| SR<br>(V/ms) | Prediction | M=7.63<br>σ=1.51<br>CDF$_{95}$=11.1 | M=11.95<br>σ=1.01<br>CDF$_{95}$=14.3 |
|  | Simulation | M=7.53<br>σ=1.48<br>CDF$_{95}$=11.0 | M=12.21<br>σ=0.99<br>CDF$_{95}$=14.5 |
|  | Error (%) | M=1.33<br>σ=2.03<br>CDF$_{95}$=-1.52 | M=-2.13<br>σ=2.02<br>CDF$_{95}$=1.49 |

| | | | |
|---|---|---|---|
| PM (degree) | Prediction | M=80.31 σ=0.057 CDF$_{95}$=80.4 | M=60.91 σ=0.066 CDF$_{95}$=61.1 |
| | Simulation | M=78.19 σ=0.061 CDF$_{95}$=78.4 | M=60.81 σ=0.071 CDF$_{95}$=61.0 |
| | Error (%) | M=2.71 σ=-6.56 CDF$_{95}$=-2.62 | M=0.16 σ=-7.04 CDF$_{95}$=0.14 |
| Run time (s.) | Prediction | < 0.01 | < 0.01 |
| | Simulation | 1350 | 1340 |

Table 3: Accuracy Verification of the Threshold Voltage at ts.

| $t_s$ (s.) | $\Delta V_t^{ts}$ | Transistors | | | | | |
|---|---|---|---|---|---|---|---|
| | | $M_3$ | $M_5$ | $M_7$ | $M_{10}$ | $M_{11}$ | $M_{13}$ |
| 10 | Golden | 0.02 | 0.25 | 0.01 | 0.38 | 2.15 | 0.06 |
| | Prediction | 0.02 | 0.25 | 0.01 | 0.38 | 2.15 | 0.06 |
| | Error (%) | 0.00 | 0.00 | 0.00 | 0.00 | 0.00 | 0.00 |
| 100 | Golden | 0.14 | 1.82 | 0.10 | 2.81 | 3.15 | 0.42 |
| | Prediction | 0.14 | 1.82 | 0.10 | 2.81 | 3.15 | 0.42 |
| | Error (%) | 0.00 | 0.00 | 0.00 | 0.00 | 0.00 | 0.00 |
| 1000 | Golden | 0.28 | 3.58 | 0.19 | 5.53 | 4.35 | 0.83 |
| | Prediction | 0.28 | 3.57 | 0.19 | 5.50 | 4.33 | 0.83 |
| | Error (%) | 0.00 | 0.28 | 0.00 | 0.54 | 0.46 | 0.00 |

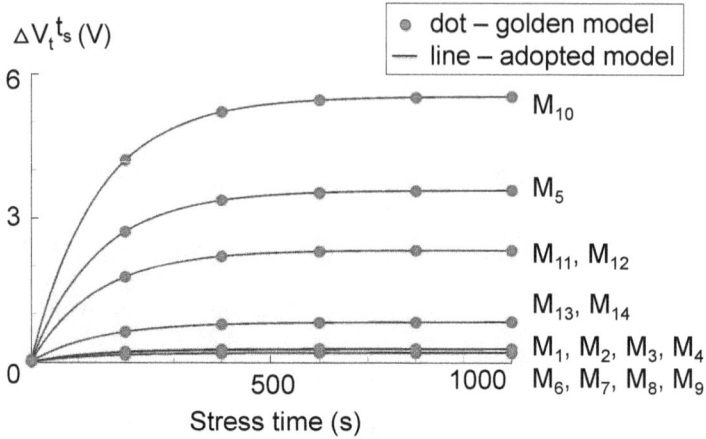

Figure 16: The threshold voltage change with degradation.

## B) Optimization Results of the OPA Circuit

For the OPA , the optimization results of different approaches are compared in Table 4. In order to make a fair comparison, these three approaches use the same bias-driven optimization engine. Only their objective functions and degradation models are different. In each experiment, the performance and design yield at $t_0$ (ts=0s) and $t_{1000}$ (ts=1000s) are both reported in Table 4. The column "$WCD_{fresh}$" shows the fresh-yield-optimized result without lifetime yield consideration. This experiment is similar to the first stage of Fig. 10. by maximizing the $WCD_{fresh}$ value. The column "$WCD_{quad}$" shows the lifetime-yield-optimized result by maximizing the $WCD_{quad}$ values at ts, which is similar to the approach proposed in the paper [19]. Quadratic degradation model is used in this approach to predict the parameter variations with aging effects. The column "CSP" shows the synthesis result of the proposed flow, which attempts to optimize the power, fresh yield and lifetime yield simultaneously with equal weights. The power consumption, the total transistor area, the design yield, and the total computation time for the synthesis procedure of each circuit are reported in the row "Overall Results".

Table 4: Optimization Results of the OPA Case.

| Performance | | WCD$_{fresh}$ | | WCD$_{quad}$ [27] | | CSP | |
|---|---|---|---|---|---|---|---|
| | | $t_0$ | $t_{1000}$ | $t_0$ | $t_{1000}$ | $t_0$ | $t_{1000}$ |
| Gain | ≥ 15 (dB) | 23.7 | 45.3 | 18.2 | 36.2 | 20.7 | 24.6 |
| GBW | ≥ 2 (kHz) | 7.4 | 0.8 | 8.2 | 2.8 | 5.6 | 2.4 |
| SR | ≥ 6 (V/ms) | 16.7 | 6.5 | 17.9 | 14.0 | 14.5 | 8.9 |
| PM | ≥ 60 (°) | 84.1 | 43.5 | 75.6 | 73.1 | 72.0 | 64.1 |
| Overall results | Power (mW) | 18.7 | | 22.8 | | 21.0 | |
| | Overhead (%) | - | | 21.8 | | 12.3 | |
| | Area (µm²) | 2940.2 | | 3789.3 | | 3021.8 | |
| | Overhead (%) | - | | 28.9 | | 2.8 | |
| | Yield (%) | 100.0 | 0.0 | 97.4 | 100.0 | 100.0 | 100.0 |
| | Run Time (s.) | < 1 | | < 1 | | < 1 | |

According to the analysis results, the original synthesis result (*WCD$_{fresh}$*) shows high fresh yield (yield=100%) at ts=0. However, the design yield becomes 0% at ts=1000s with significant degradation effects. By using the results of *WCD$_{fresh}$* as the initial solutions, the optimization results with lifetime yield consideration (*WCD$_{quad}$*) can keep high design yield at ts=1000s. However, this two-stage approach does not simultaneously consider the fresh yield, life-time yield, and design overhead. The overhead compared to the design without lifetime yield consideration is large, as shown in Table 4. In addition, the fresh yield of the circuit generated by this approach is no longer optimal.

The results of the proposed flow (CSP) reach the highest lifetime yield with the lowest overhead. Each performance with degradation is much closer to the given specifications. It shows that the adopted accurate degradation model helps to predict accurate performance and reduce the over-design. Most importantly, both the fresh yield and lifetime yield are optimized simultaneously. Taking the Gain and GBW of the OPA circuit as examples, the performance distribution of the Monte Carlo analysis shown in Fig. 17 also demonstrates this trend of yield improvement.

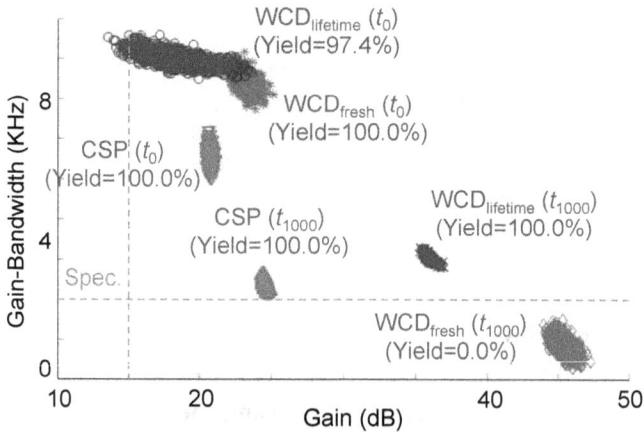

Figure 17: Performance distribution of the OPA case.

## C) Optimization Results of the DAC Circuit

In this section, the 4-bit DAC [8] is used as another example to demonstrate the proposed design flow. The schematic of DAC is shown in Fig. 18. The key performances are integral non-linearity (INL), differential non-linearity (DNL) and settling time. According to the experimental results in Table 5, the same trend as the one in the OPA case can be observed. The original synthesis result ($WCD_{fresh}$) shows high fresh yield (yield=100%) at ts=0. However, the lifetime yield becomes 0% with transistor degradation. The experimental results of $WCD_{quad}$ show a high lifetime yield (yield=100%) as ts=1000. However, the fresh yield is not optimal due to the two-stage optimization approach. The optimal results of the proposed design flow (CSP) reaches the highest fresh yield and lifetime yield with the lowest design over-head due to the simultaneous consideration of fresh yield and lifetime yield. Taking the settle time and DNL of the DAC circuit as examples, the performance distribution of the Monte Carlo analysis shown in Fig. 19 also demonstrates this trend of yield improvement.

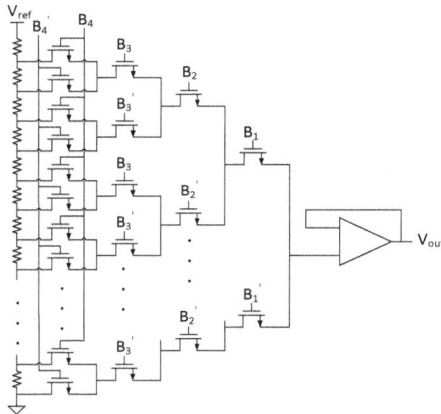

Figure 18:  The schematic of DAC circuit.

Table 5: Optimization Results of the DAC Case.

| Performance | | $WCD_{fresh}$ | | $WCD_{quad}$ [27] | | CSP | |
|---|---|---|---|---|---|---|---|
| | | $t_0$ | $t_{1000}$ | $t_0$ | $t_{1000}$ | $t_0$ | $t_{1000}$ |
| Settling time | < 100 (us) | 50.7 | 101.3 | 55.2 | 77.3 | 52.3 | 86.3 |
| DNL | < \|0.5\| (LSB) | 0.13 | 0.53 | 0.45 | 0.17 | 0.23 | 0.30 |
| INL | < \|1.0\| (LSB) | 0.28 | 0.77 | 0.63 | 0.31 | 0.32 | 0.45 |
| Overall results | Power (mW) | 26.9 | | 35.3 | | 29.3 | |
| | Overhead (%) | - | | 31.2 | | 8.9 | |
| | Area ($\mu m^2$) | 6703.4 | | 8661.3 | | 6897.1 | |
| | Overhead (%) | - | | 29.2 | | 2.9 | |
| | Yield (%) | 100.0 | 0.0 | 97.0 | 100.0 | 100.0 | 100.0 |
| | Run Time (s.) | < 1 | | < 1 | | < 1 | |

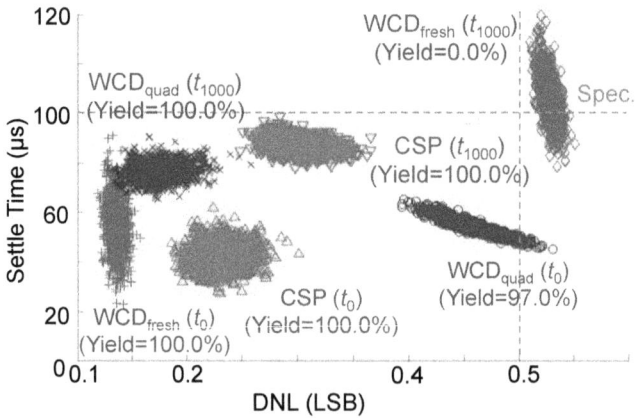

Figure 19: Performance distribution of the DAC case.

## 5. Conclusion

In this chapter, two automated robust design optimization techniques are introduced for the analog circuits with flexible TFTs. An incremental Latin hypercube sampling approach is first introduced to analyze the aging effects of analog circuits. By reusing previous sampling information, only a small portion of samples are incrementally updated in different time steps, which significantly reduces the number of simulations for aging analysis. Compared to the MC results, 243x average speedup can be obtained by this approach while keeping the accuracy of PDF estimation.

Second, an automatic robust optimization technique is introduced for the analog circuits to deal with the severe parameter variations and critical degradation effects in flexible electronics. A new yield evaluation term, CSP, is integrated into the equation-based sizing approach to optimize the fresh yield and lifetime yield at the same time with accurate variation consideration. A simple transformation method is also proposed to avoid the rebuilding cost of transistor models at different times. Compared with optimizing the two yield issues separately, the proposed simultaneous optimization approach can improve the fresh yield and lifetime yield at the same time with a great reduction on area and power overhead, as shown in the experiments. Those results demonstrate that the proposed approach is able to help designers solve the main difficulty of designing the analog circuits in flexible electronics.

# References

[1] T.-C. Huang, J.-L. Huang, and K.-T. Cheng, (2011). Robust Circuit Design for Flexible Electronics. *IEEE Design & Test of Computers.* 28(6), 8-15.

[2] A. Singhee and R. Rutenbar, (Mar. 2007). From finance to flip flops: A study of fast quasi-monte carlo methods from computational finance applied to statistical circuit analysis. *Int'l Symp. on Quality Electronic Design.* 685-692.

[3] J. Jaffari and M. Anis, (Jan. 2011). On efficient LHS-based yield analysis of analog circuits. *IEEE Trans. on Computer-Aided Design.* 30(1), 159-163.

[4] J. H. Cheon, J. H. Bae, and J. Jang, (Mar.2008). Mechanical stability of poly-Si TFT on metal foil. *Solid-State Electron.* 52(3), 473–477.

[5] J. N. Burghartz, N. Wacker, M.-U. Hassan, H. Rempp, and H. Richter, (2009). MOST modeling for ultra-thin flexible electronics. *MOS-AK Workshop, MOS Modeling and Parameter Extraction Working Group.*

[6] R. Shringarpure, S. Venugopal, L. T. Clark, D. R. Allee, and E. Bawolek, (Jan.2008). Localization of gate bias induced threshold voltage degradation in a-Si:H TFTs. *IEEE Electron Device Lett.* 29(1), 93–95.

[7] S.-E. Liu, C.-P. Kung, and J. Hou, (Jan. 2009). Estimate threshold voltage shift in a-si:h tfts under increasing bias stress. *Electron Devices, IEEE Transactions on.* 56(1),65–69.

[8] Y.-C. Tarn, P.-C. Ku, H.-H. Hsieh, and L.-H. Lu, (May.2010). An amorphous-silicon operational amplifier and its application to a 4-bit digital-to-analog converter. *Solid-State Circuits, IEEE Journal of.* 45(5), 1028–1035.

[9] X. Pan and H. Graeb, (Sep. 2010). Reliability analysis of analog circuits using quadratic lifetime worst-case distance prediction. *IEEE, Custom Integrated Circuits Conference.* 1–4.

[10] Y.-L. Chen, W. Wu, C.-N. J. Liu, and L. He,( Jan. 2015). Incremental Latin Hypercube Sampling for Lifetime Stochastic Behavioral Modeling of Analog Circuits. *Asia and South Pacific Design Automation Conference.*

[11] Y.-L. Chen, W.-R. Wu, C.-N. J. Liu, and J. C.-M. Li, (Jan. 2014). Simultaneous Optimization of Analog Circuits with Reliability and Variability for the Applications on Flexible Electronics. *IEEE Trans. on Computer-Aided Design.* 33(1), 24-35.

[12] T. McConaghy and G. Gielen, (Nov. 2009). Globally reliable variation-aware sizing of analog integrated circuits via response surfaces and structural homotopy. *IEEE Trans. Comput.-Aided Design,* 28(11), 1627–1640.

[13] W. Gao and R. Hornsey, (2010). A power optimization method for CMOS op-amps using sub-space based geometric programming. *Proc. Design Test Eur.* 508–513.

[14] X. Li, J. Wang, L. T. Pileggi, T.-S. Chen, and W. Chiang, (2007). Performance-centering optimization for system-level analog design exploration. *Proc. Int. Conf. Comput.-Aided Design.* 464–469.

[15] S. Deyati and P. Mandal, (2011). An automated design methodology for yieldaware analog circuit synthesis in submicron technology. *Proc. Int. Symp. Quality Electron. Design.* 1–7.

[16] H. E. Graeb, (2007). Analog Design Centering and Sizing. *Springer.*

[17] K. J. Antreich, H. E. Graeb, and C. U. Wieser, (2000). Wicked: Analog circuit synthesis incorporating mismatch. *Proc. Custom Integr. Circuits Conf.* 511–514.

[18] R. Shringarpure, S. Venugopal, Z. Li, L. T. Clark, D. R. Allee, E. Bawolek, and D. Toy, (Jul. 2007). Circuit simulation of threshold-

voltage degradation in a-Si: H TFTs fabricated at 175 °C. *IEEE Trans. Electron Devices.* 54(7), 1781–1783.

[19] X. Pan and H. Graeb, (Aug. 2012). Reliability optimization of analog integrated circuits considering the trade-off between lifetime and area. *Microelectron. Reliab.*, Elsevier. 52(8), 1559–1564.

[20] X. Pan and H. Graeb, (Feb. 2011). Lifetime yield optimization of analog circuits considering process variations and parameter degradations. *Advances in Analog Circuits.* InTech.

[21] Y.-F. Cheng, L.-Y. Chan, Y.-L. Chen, Y.-C. Liao, and C.-N. J. Liu, (Jul. 2012). A Bias-Driven Approach to Improve the Efficiency of Automatic Design Optimization for CMOS OP-Amps. *Proceedings Asia Symposiums on Quality Electronic Design*, 59-63.

[22] F. Liu and S. Ozev, (Jan. 2005). Hierarchical Analysis of Process Variation for Mixed-Signal Systems. *Proceedings IEEE Asia and South Pacific Design Automation Conference (ASP-DAC)*, 465–470.

# CHAPTER THREE

## Inorganic Semiconducting Nanowires for Flexible Electronics

D. Shakthivel, C. García Núñez, and R. Dahiya

*Bendable Electronics and Sensing Technologies (BEST) Group, Electronics and Nanoscale Engineering (ENE), School of Engineering, University of Glasgow, G12 8QQ, UK*

## 1. Introduction

Silicon based CMOS integrated circuit (IC) technology has dominated electronics industry for the past 50 years [1]. Growth in this domain revolutionized other areas such as computation, communication and information technology. In Si technology, a solid rigid substrate of thickness in the order of 100s of microns supports the planar devices and circuits. This makes all electronic systems bound to be rigid. Many of the futuristic research areas such as photovoltaic, display, sensing and wearable electronic technologies demands the requirement of flexibility and large area functional electronics [2, 3]. For example, a large area conformable touch sensor assembly could be useful in the fabrication of electronic skin (*e*-skin) for robotics and prosthetics [4, 5]. One of the successful approaches were to thin-down the Si IC chips up to tens of microns to enhance flexibility [6]. However, attributes such as reversible deformability, sharp curvatures, high elasticity, resistance to crack, cost effective large area manufacturability are great incentives to use the CMOS technology for flexible electronics. An alternative approach is to fabricate electronics over flexible polymer substrates (plastics, Polyethylene terephthalate (PET), Poly (methyl methacrylate) (PMMA), polyimide, etc.,) to meet the application demands [7-10]. The major issue is to overcome the high device processing temperatures. One of the classical examples was the successful fabrication of Si based thin film solar cells over flexible polymer substrates [11]. This technology shows the potential behind the large area flexible electronics. Active flexible applications including paper displays and wearable computers have been developed so far [12]. Flexible devices based on hydrogenated amorphous silicon (a-Si:H) [13], and organic semiconductors [14] have been investigated. Similarly, Si thin film transistors over polymer substrates have shown good potential in flexible large area displays.

In summary, the real challenge is to utilize the Si device fabrication technology over flexible polymer substrates. This opens the avenue of heterogeneous functional device integration over flexible materials [15]. Advantageously, this approach helps to utilize wide variety of materials systems for different applications. In case of CMOS, functionality is achieved through only silicon planar devices. The evolution of nanomaterials combined with device fabrication and integration technologies shows good promise in this area. The high functionality of nanomaterials due to properties such as large flexibility, high aspect ratio, and low dimensions, make them very attractive for the development of high performance electronics over flexible substrates, which could be potentially used in applications including biosensors, photodetectors, photovoltaics, etc. In recent years, nanowires (NWs) have shown to be the gripping class of materials to be used as an active channel for nanoelectronic devices. The high aspect ratio of these nanomaterials makes them highly sensitivity to the environment which is promising for the development of sensors capable of detecting different kinds of stimulus, comprising, gases, liquids, light, etc.

## 2. Semiconducting Nanowires

NWs are one-dimensional (1D) anisotropic structures in which two of the wire's dimensions are restricted to less than 100 nm with the third one extends up to several tens of microns [16-18]. NWs of elemental, compound and ternary alloys have been synthesised using many different physical and chemical methods. Inorganic materials including IV (Si, Ge), III-V (GaAs, InAs, GaN, GaP), II-VI (ZnO) semiconductors, could be prepared in NW morphologies with diameters down to sub ten nanometers. Special characteristics of these NWs include their aptness for nanoelectronic device fabrication, sensitive to wide optical wavelengths in the electromagnetic spectrum, nanoscale mechanical flexibility and sharp response to chemical to bio molecules due to high surface to volume ratio. These 1D structures outperform their bulk and thin film counterparts due to their single crystalline nature and controlled chemical composition [19]. Importantly, NWs exhibit interesting electronic properties if the diameter reduces less than excitation Bohr radius of the material [20]. Carbon nanotubes (CNTs) are one of the widely explored 1D nanostructures for widespread applications such as nano field effect emitters (FETs), chemical and biological sensors [21, 22]. Similarly elemental and compound semiconductor NWs have been investigated for many attractive applications. For example, the hole mobility in Ge NWs was found to be 5 times higher than bulk Si which is suitable for NW HEMTs. Materials such as ZnO, GaN, InN have been underused due to the unavailability of bulk wafers and difficulty in producing them as a thin film epitaxial structures [23, 24]. ZnO is a wide band gap ($E_g$ = 3.4 eV) semiconductor material with high electrical conductance and low-cost fabrication process which are attractive characteristics for the development of ultraviolet (UV) optoelectronic devices for applications such as flame sensing, UV radiation calibration and monitoring, chemical and biological analysis, space research, environmental monitoring, and optical communications. An ideal UV photodetector should be very sensitive in the UV region with high signal-to-noise ratio, high selectivity, and fast response. For that reason, ZnO nanostructures have received considerable attention in the past years due to their good electrical and mechanical properties, which along their high surface-to-volume ratio allows improving the performance of devices such as optical switches, light, biological and gas sensors [25-27]. The growth mechanisms of III-V NWs such as GaAs NWs have been thoroughly investigated [28, 29], in order to obtain single crystal nanostructures that allow to develop different applications such as nanometric solar cells and high-speed electronics.

NWs are successful candidates in preparing devices and structures such as logic and memory, sensors (chemical, bio, gas, etc.,), photovoltaics, NW light emitter diodes (LEDs), and lasers. Additionally, a variety of NW based heterostructures could be realized using current chemical methods[30]. Periodically stacked segments of different semiconductor NWs finds interesting applications in resonant tunneling diodes [31] and single electron transistors [32]. These axial heterostructures have been demonstrated for materials system such as Si/Ge/Si, GaP/GaAs etc., Segment periodicity, length and interfacial abruptness are important for different applications [33]. Similarly, core-shell heterostructures are another interesting class of NW variants which offers interesting electrical and optical properties. For instance, core-shell structures enhance 1D electron and hole gases which have been used to show NW

HEMT devices [34]. Core-multi shell structures allow band gap engineered interfaces which help to control the charge confinement at the interfaces. Semiconducting materials such as SiGe and GaN/AlGaN structures were also shown to exhibit these phenomena.

## 3. NW Synthesis Methods

Semiconducting NWs have been shown to synthesize a variety of approaches [35-40]. Largely, these approaches have been divided into two categories depends on the source chemistry. First one utilizes the bulk material of the NW which were reduced into arrays of NWs. Si NWs produced from the single crystalline bulk wafer is the proven example of this approach. The chemical composition of the NW is decided by the initial wafer material. The second approach usually begins with the atoms/molecules/organic complexes and allow them to react in a controlled ambient to build the NWs from bottom-up. Bottom-up methods provide sufficient freedom to tune structure, dimension, spacing and composition of the NW array. Both approaches utilize a wide variety of physical and chemical methods in the NW synthesis process.

### 3.1 Bottom up-Method
Vapor-liquid-solid (VLS) mechanism (Fig.1) plays a crucial role semiconducting NWs research [41-43]. NWs of lower bound diameter 3 nm to 10s of micron have been grown by VLS mechanism. One of the biggest advantageous of this method is the ability to grow NWs of elementary, binary and ternary semiconducting materials [38]. VLS mechanism uses a metallic catalyst particles for the synthesis of NWs [18]. Preparation of array of catalyst particles is the first step in the synthesis process.

NW diameter, site and interspacing of the NWs are decided by the catalyst particle preparation method. After the growth of NWs, catalyst particles are firmly attached at the tip of the NWs. A variety of physical and chemical techniques such as sputtering, pulsed laser deposition, MBE, CVD, etc., have been used to execute the VLS mechanism. Among the techniques, CVD offers greater versatility, atomic level control in source supply and a wide range of source materials.

Figure 1: Illustration of bottom-up VLS growth mechanism. (a) Schematic diagram of the VLS process, (b) SEM image of Au catalyst prepared over Si substrate, (c) SEM cross sectional image of Si NWs.

Additionally, chemical vapor deposition (CVD) is temperature enhanced dry chemical technique which is compatible with Si CMOS fabrication technology. The majority of the NW device prototypes have been demonstrated by using CVD method. CVD assisted VLS growth mechanism provides control over NW dimension, morphology (axial, core-shell heterostructures and branched), site and dopant concentration. Hence it is important to understand the atomistic aspects of VLS mechanism (Fig.2 (a)) to make use of its full potential of NWs [44-52]. Kinetics of the of the VLS growth mechanism is reported through various models to elucidate the significance of all the atomistic processes.

For the case of Si NW growth, catalysts such as Au, Al, Pt, Cu, Fe etc., have been used with suitable temperatures[50].

Suitability of the catalyst particles is decided by the ability of the two elements mix at elevated temperatures. Metallurgical binary phase diagrams could be used for this purpose to choose the appropriate catalyst for a particular NW material synthesis. For example, Au is a commonly used catalyst for the growth of Si NWs. Au forms a liquid solution with 19atom% Si at and above the temperature of 360°C. All the catalyst particles must possess this eutectic point feature at a specific temperature where a solid-liquid transition occurs. This is the minimum temperature at which the NW growth experiment could be performed. In a typical Si NW growth process, Au nano particles are placed in an inert ambient at a temperature about ~500°C. Si source molecules ($SiCl_4$ or $SiH_4$) are introduced and allowed to adsorb over Au catalyst particles to inject Si into them. Si incorporation leads to the formation of Au-Si liquid solution which eventually saturated by Si. This set up the Si NW precipitation at the catalyst-substrate interface with a diameter equal to that of the catalyst particle. Si concentration in the catalyst particle, nature of the source gases, growth temperature are the dictating parameters to obtain the electronic quality NWs. To date, none of the *in-situ* and *ex-situ* experimental tools were successful in measuring the Si concentration prevailing in the catalyst particles during NW growth. Analysis of the kinetics of the VLS growth process provides more insights into the atomistic processes involved [53].

The thermodynamic driving force for NW growth from supersaturated catalyst-NW materials liquid solution is given by,

$$\Delta\mu_{LS} = kT\ln(\frac{C}{C^o}) \qquad (1)$$

where, $C$ is the concentration (in atoms/m$^3$) of Si in the Au droplet, $C^o$ is the equilibrium concentration which can be obtained from the phase diagram.

NW growth process governed by five atomistic processes. They are, (1) adsorption at the vapor/catalyst interface, (2) Surface diffusion of Si from NW sidewall to the catalyst particle, (3) Si evaporation from the catalyst particle, (4) Si escape from the catalyst through reverse reaction, (5) Crystallization of Si NW at the interface. Here, process (1-2) are Si injection into the catalyst and (3-5) corresponds to ejection out of the droplet. At steady state, a balance between injection and ejection which results in a Si concentration in the droplet. Hence the VLS steady state is expressed as,

$$\left[\frac{dC}{dt}\right]_{Injection} = \left[\frac{dC}{dt}\right]_{ejection} \qquad (2)$$

where, $C$ is the concentration of atoms (in number per unit volume) in the catalyst droplet and $t$ represents time. By including all the injection and ejection steps, Eq.2 can be written as:

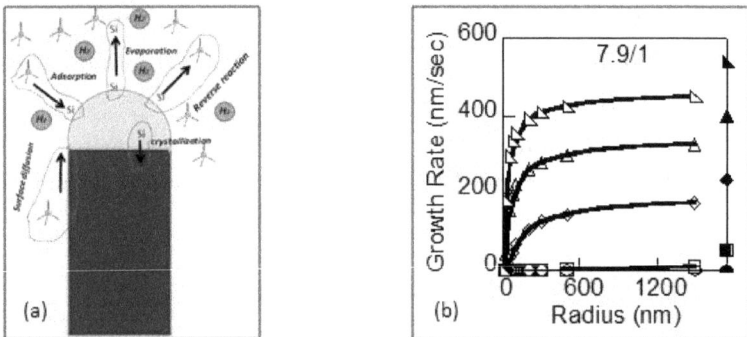

Figure 2: (a) Schematic representation of the steady state growth of Si

NWs by VLS mechanism incorporates all the atomistic processes. (b) Estimated growth of Si NWs with respect to the NW diameter.

$$\left[\frac{dC}{dt}\right]_{Adsorption} + \left[\frac{dC}{dt}\right]_{surfacedifusion} =$$

$$\left[\frac{dC}{dt}\right]_{evaporation} + \left[\frac{dC}{dt}\right]_{reverse} + \left[\frac{dC}{dt}\right]_{NWgrowth} \quad (3)$$

At steady state, the net injection rate is equal to the rate of crystallization of NW at the interface. By balancing injection and ejection using suitable expressions, one arrives to estimate the steady state Si concentration of the droplet. The NW growth rate is directly proportional to the Si concentration in the droplet. Crystallization at the interface occurs through predominantly layer birth and spread mode or layer by layer (LL) mode [54]. NW growth rate through this mode is expressed as follows [55],

$$\left(\frac{dh}{dt}\right)_{LL-IF} = (J\pi R^2)a \quad (4)$$

where, $R$ is wire radius, $a$ is the layer step height, and $J$ is a number of nuclei formed per unit area. NW growth rate is estimated by using this method and found to be fall within an order of experimentally measured values. The growth rate dependency over temperature and Si source pressure has been identified through the kinetic study. This provides a road map to identify the experimental conditions for the NW synthesis (Fig.2 (b)). The temperature dependency is the key as it clearly defines the required source precursors and catalyst particles suitable for the required NWs.

## 4. Nanowire Transfer Methods

The use of inorganic semiconductor NWs in flexible electronics has been extended during the last two decades mainly due to their high intrinsic flexibility and excellent optoelectronic properties [56-58]. As mentioned in the previous section, the growth of NWs via either bottom-up method required the use of extremely high temperatures which are not compatible with the direct growth on flexible substrates. However, this range of temperatures is completely necessary to obtain a high crystal quality structure necessary to fabricate high performance electronic devices. In this regard, different transfer methods have been investigated aiming to design and to fabricate flexible electronics based on NWs [59]. The investigation of different transfer methods comprises the control over the density of assembled NWs, their alignment, and uniformity, as well as, the preservation of their as-grown characteristics. The lack of these factors would hinder the high performance of the resultant flexible electronics. Moreover, the development of this promising research aims to obtain a reliable and scalable NW assembly technique, which allows fabricating large-area flexible electronics.

Semiconductor NWs have been transferred from a growth substrate to a foreign substrate, including both rigid and flexible substrates, using different techniques. In general, there are two main ways to transfer NWs into a flexible substrate, obtaining a high assembling yield, comprising transfer printing and solution based methods. Transfer printing (Fig.3) consists in a directionally sliding of a donor substrate with randomly grown NWs over a receiver substrate, leading to an assembly of aligned parallel arrays of NWs [59-61]. Since the NWs are brought into direct contact with the surface of the receiver substrate, during the sliding process, the shear force tends to re-align the NWs along the sliding direction. Nature of the NW-donor interface and pressure between both donor and receiver substrate are the key parameters to promote the NW detachment. In order to minimize the NW-NW interaction, as well as, to preserve the initial characteristics of the NWs, i.e. structural and morphological properties, the coating of the receiver substrate with lubricants was demonstrated to extremely improve the NW transfer yield [59]. Furthermore, surface functionalization of the receiver substrate has showed fruitful results in terms of NW alignment over large areas up to 4 inch Si wafers, and high linear densities (~ 7 NW/□m) [59], latter being comparable to solution based techniques such as Langmuir-Blodgett (LB) [62, 63] and much larger than bubble-blown technique [64] . The high NW density obtained by contact printing technique is very attractive for the development of single NW based electronic devices, as well as, high electrical current devices based on a controlled number of aligned NWs. In addition, the use of patterned receiver substrates allows creating more complex structures via hierarchical printing procedures, resulting in crossed NWs and controlled staking of the NWs.

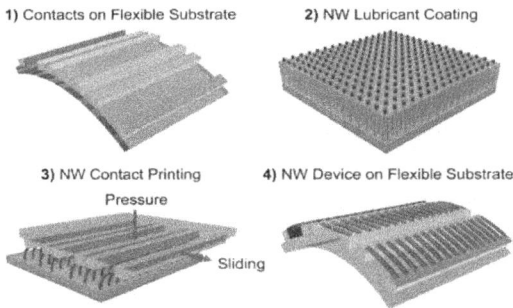

Figure 3: Illustration of the contact printing procedure, including (1) Electrode patterning on a flexible substrate, (2) Chemical treatment of the NW sample (3) Contact printing process (4) Device structure.

Contact printing approach was further extended to the fabrication of flexible electronics, carrying out the same procedure but in this case using a cylindrical shape donor substrate with NWs. This technique is known as roll-printing (Figure 4) and has shown good compatibility with almost all the typical used substrates, including rigid Si, glass, quartz, flexible substrates such as plastic [65].

Figure 4. 3D schematic illustration of the roll-printing fabrication procedure of flexible electronics based on NWs.

Tape pealing is a low-cost fabrication procedure consisting in the use of an adhesive tape to remove NWs from the donor substrate and, after that, stamp them onto a receiver substrate with pre-patterned electrodes (Figure 5).

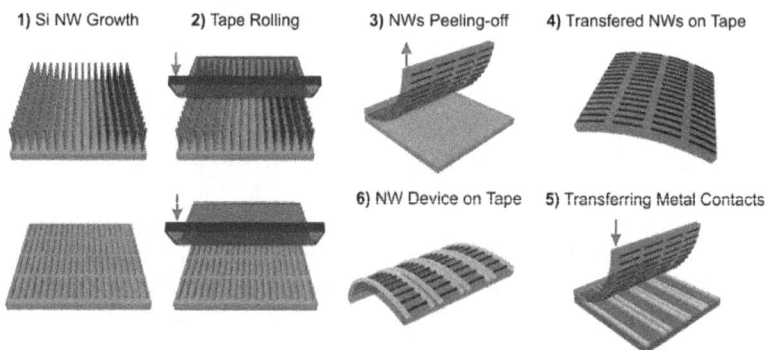

1) Si NW Growth    2) Tape Rolling    3) NWs Peeling-off    4) Transfered NWs on Tape

6) NW Device on Tape    5) Transferring Metal Contacts

Figure 5: Schematic illustration of the tape peeling fabrication procedure. (1) NWs vertically and horizontally aligned on a rigid substrate; (b) Tape rolling of an adhesive flexible substrate; (3) Peeling of the NWs; (4) NWs stamped on the adhesive tape; (5) Stamping of the NWs from the donor tape to the receiver substrate with patterned electrodes; (6) Resultant devices.

An alternative way to integrate NWs in flexible substrates is based on solution methods. Typically, NWs are suspended in an organic solvent and later dispersed on a receiver substrate. In this regard, drop-casting of NWs on top of pre-patterned metallic electrodes is probably a low-cost method; however, the obtained low yield make researchers to further investigate other alternatives such as Langmuir Blodgett (LB) [62, 63], bubble-blown techniques [64], electric field alignment [58, 66]. LB technique is used for the alignment of nanostructures floating onto the surface of a liquid. To carry out the assembly of NWs via LB technique, it is necessary to functionalize NW surface in order to allow them float on the liquid surface like "logs on a river" (Fig.6 (a)). Then, the compression of the liquid with an appropriate barrier leads to the NW alignment preferentially perpendicular to the compression direction (Fig.6(b)). Finally, a receiver substrate can be immersed in the solution and extracted in a controlled way, receiving NWs already aligned along its surface (Fig.6(c)). Speeds of both substrate extraction and the horizontal barrier will determine important parameters such as NW center-to-center spacing, NW density, etc. LB technique has been successfully used to develop the assembly of a different kind of NWs, including Si [67], Ag [68], Ge [69]. As in the case of contact printing, LB also allows carrying out hierarchical steps resulting in complex devices. This technique makes possible the assembly of a high density of NWs comparable to this obtained by contact printing, and is completely compatible with rigid and flexible substrates. Moreover, LB has shown a certain selectivity to the size of the assembled NW, showing better successful rates on NWs with diameters ranged between 20-40 nm than those below 15 nm [70].

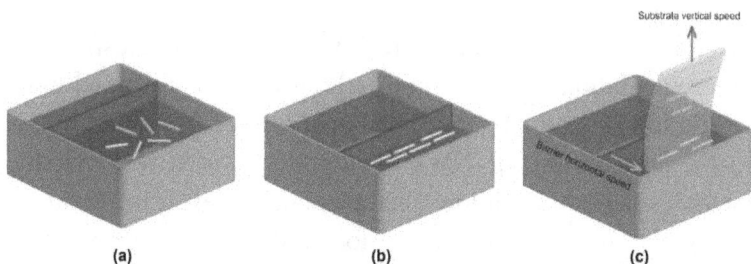

(a)                          (b)                          (c)

Figure 6. Schematic illustration of the LB method to align NWs on a flexible substrate. (a) Functionalized NWs floating on the liquid surface; (b) Compression makes NWs aligned; (c) Transferring of the NWs to the flexible substrate.

In the bubble-blown technique, NWs are embedded in a polymeric suspension which is expanded forming a bubble (Fig. 7a). The shear force created by the controlled expansion of the bubble promotes the alignment of the NWs along the bubble surface. Thereafter, bubble walls are brought into contact with the receiver substrate, transferring the aligned NWs to the substrate. One of the most important advantages of this technique overall is its potential use as large-area integration technique, showing high transferring yields on plastic sheets of $225 \times 300$ mm$^2$ and Si wafers of 200 mm in diameter. However, the main disadvantage of the bubble-blow is the low NW density around 0.04 NW/$\square$m$^2$ that can limit the functionality of this technique for the fabrication of high current output devices at the nanoscale. Flow assisted technique (Fig. 7b) align the NWs in the direction of fluid flow using shear force in the fluid/NW boundaries. The accurate manipulation of NWs suspended in a liquid can be also carried out using electric fields to align and to trap them at preferential sites. In this case, suspended NWs tend to be re-aligned along the fluid direction due to the effect of the shear force created by the fluid, minimizing the drag force exerted on the NW surface. The main disadvantage of this technique is the assembly area is strongly limited by the cross-section of the microchannel conducting the fluid which is in the micrometric range. Although the alignment length that can be achieved with flow-assisted technique is observed in the mm scale. This technique also has the functionality to fabricate complex structures including crossed NW arrays by alternating the flow direction through different step assembly processes. These processes can be carrying without affecting alternating steps, which allows fabricating homo- and hetero-structures composed by NWs of different types.

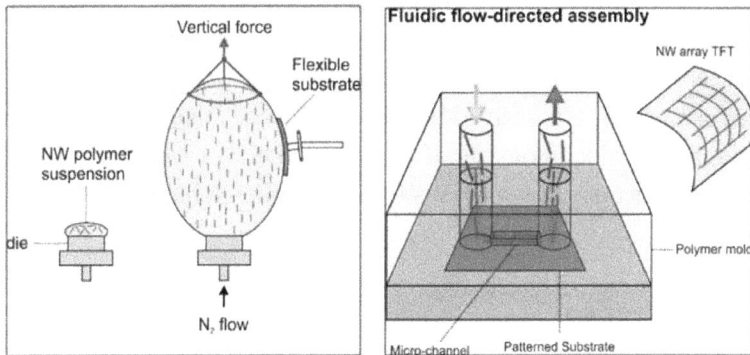

Figure 7: (Left) Schematic illustration of the BB method to align NWs. (Right) Fluidic flow assembly of NWs on a flexible substrate.

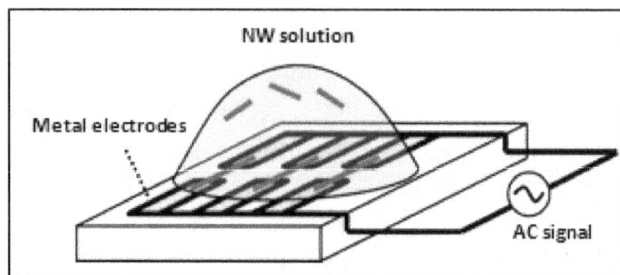

Figure 8: Schematic illustration of DEP NW assembling process between the electrodes

Dielectrophoresis (DEP) is another well-known solution based technique extensively used to assemble NWs on different kind of substrates. The technique typically uses either non-uniform electric fields to polarize suspended NWs, attracting them to specific places on the sample (Figure 8). In this regard, factors such as electrical properties of both the NW and the liquid medium, NW morphology, and conditions of the non-electric field are the main parameters controlling the performance of the assembling process [27, 58, 66, 71]. DEP presents an excellent control over the assembly yield, and NW alignment, as well as, preventing the use of mechanical forces that can damage the nanostructure. Recently, DEP has been combined with fluidic alignment, which consists of a standard DEP process but carried out on a laminar flow of NWs through a micro channel [58].

## 5. Applications

Many interesting applications have been envisaged for the heterogeneous integration of NWs over flexible organic materials [59, 72-76]. Despite the tremendous developments in the NW synthesis has been achieved, NW transfer processes have not matured enough for industrial manufacturing. However, many system level applications have been demonstrated in the past decade.

## 5.1 Nanowire FETs over Flexible Materials

NW based FETs are the core of future nanoelectronics for logic and memory circuits[74]. Back gated and wrap gated FETs have shown to compete with the Si based planar electronics. Importantly, excellent control over the charge carrier in the active NW channel is obtained by wrapping the gate oxide around the NWs. NW FETs are building blocks for the most of the envisaged applications in NW based flexible electronics. Flexibility and high mobility are the key factors to compete with the organic semiconductors to produce high performance bendable electronics. p-type Si NWs of 20 nm diameter assembled over plastics for the fabrication of FETs using $SiO_2$ as gate dielectric and Pd as source-drain contacts. All the device fabrication steps were carried out in room-temperature due to the inherent substrate material limitations. From the $I\text{-}V$ characteristics, the estimated mobility was ~200 $cm^2/V$ s which is more than order compared with organic semiconductors for flexible electronics. NW device contacts over flexible substrates were found to be ohmic and the device characteristics are unaffected with respect to bending. These studies qualify the inorganic NW arrays for the logic gate structures with low power consumption (~1 V). Additionally, NWs of III-V semiconductors such as InAs, GaN, etc., have been explored to utilize their direct wide band gap properties. Moreover, their ability to perform at higher bending radii, great carrier mobility and innate mechanical strength are added advantageous in flexible electronic devices. For example, InAs NW FETs have been observed to operate in GHz frequency range which could be used to fabricate high frequency RF amplifiers. Similarly, GaN NW devices over flexible substrate lead to the demonstration of bendable LEDs. This could be exploited in the future for flexible displays which could be far superior compared to the organic LED/polycrystalline system.

## 5.2  Flexible Sensors and Systems

A bendable sensing system is ideal for many applications such as biomedical, gas (toxic, flammable.,), tactile sensors, optical and temperature sensing due to the conformable coverage over the subject. The key factors are sensitivity and response time could be greatly enhanced by high surface inorganic NWs. Additionally, sensing and signal processing could be easily integrated within a single substrate material. Si NW FETs prepared using lithographic top-down approaches followed by super lattice NW pattern transfer process have found to be efficient chemical sensor [77]. The adsorption based vapor sensor array is tested for $NO_2$ gas which is commonly emitted from automobiles. The sensor system responded with ~3000% increase in current for the exposure of 20 p.p.m of $NO_2$. The sensor was observed to sensitive up to the lower limit concentration of 20 p.p.b where the safe exposure limit is 53 p.p.b. Inherent high surface to volume ratio of the Si NWs enhances high sensitivity and low noise level during operation. The recovery of the exposed surface could be done easily using simple gas flush as well as low temperature treatments. The NWs array based chemical sensor over plastics is envisaged to produce artificial nose in the future. Si surface could be functionalized using a variety of organic molecules for the bio-sensor applications.

Flexible tactile sensors are important an area for applications in robotics and prosthetics. NW sensing element over flexible materials provides a natural solution to produce artificial skin (or e-skin). Javey group [56, 65] have demonstrated a large area (7x7 cm$^2$) e-skin using Si/Ge core shell NWs over flexible polyimide substrate. They have adopted the contact printing transfer process to produce a uniform array of CVD synthesized NWs over the flexible substrate. A pressure sensitive rubber (PSR) is connecting the NW FETs with their metal electrodes. An external pressure changes the conductance of the PSR which in turn modulates the FET channel conductivity. The performance of the system is investigated under different bending and rolling conditions. Low operating voltage, high carrier mobility and mechanical flexibility makes the NW based artificial skin far superior compared to the organic based systems. In compound semiconductors, ZnO NW based piezotronic transduction mechanisms have been tried for tactile sensing devices [78]. This transfer free method utilized as grown (vertical) ZnO NWs in the metal-semiconductor-metal configuration as a touch sensor. The polarization induced mechanism is capable of sensing the pressure between few kPa to 30 kPa, which falls well within the range of human skin.

## 6. Conclusions

This chapter presented the suitability and advantages of semiconducting NWs for flexible and large area electronics. Various aspects of NWs such as physical and chemical properties, synthesis methods, processing, and applications are reviewed with examples. Importantly, the electrical characteristics and device performance of NWs are superior in comparison with the state of the art organic based electronics.

However, the processing techniques need to be established for NW based applications such as electronic skin, wearable electronics, flexible optoelectronics etc. These demands feed plenty of research interest in NWs based flexible and large area electronics in academia and industry. NW synthesis process sets the platform as it dictates dimensional control, structure (axial and core-shell), composition, intrinsic stress, density and yield. NW growth mechanism and execution technique could be chosen depending on the application requirement. Many growth techniques developed in the past and their growth mechanisms have been studied using experiments and theoretical methods. Among them, catalyst particle enhanced VLS mechanism stands tall due to its versatility in application and atomic level control. Processing the synthesized NWs for flexible electronics poses a great challenge in comparison with organic materials. The problem arises due to the lack of growth techniques to produce NWs directly over flexible polymers. Controlled monolayer assembly, high density, alignment and uniform interspacing are important parameters to fabricate promising flexible circuits. NW transfer processes over flexible substrates being studied extensively to extend the demonstrated applications for industrial production. Flexible device prototypes have been demonstrated in the areas of logic, memory, sensors, energy, optoelectronics, electrodes and display. The performance of these systems is above par with rigid planar systems. However, promising scalable fabrication strategies need to be developed for commercialization.

## Acknowledgement

This work was supported by Engineering and Physical Sciences Council (EPSRC) Fellowship for Growth – Printable Tactile Skin (EP/M002527/1).

# References

[1] Moore, G.E.: 'Cramming More Components Onto Integrated Circuits', Proceedings of the IEEE, 1998, 86, (1), pp. 82-85

[2] Dahiya, R.S.: 'Epidermial electronics: flexible electronics for biomedical applications', 2015

[3] Wong, W.S., and Salleo, A.: 'Flexible electronics: materials and applications' (Springer Science & Business Media, 2009. 2009)

[4] Dahiya, R.S., Mittendorfer, P., Valle, M., Cheng, G., and Lumelsky, V.J.: 'Directions toward effective utilization of tactile skin: A review', Sensors Journal, IEEE, 2013, 13, (11), pp. 4121-4138

[5] Dahiya, R.S., and Valle, M.: 'Robotic tactile sensing: technologies and system' (Springer Science & Business Media, 2012. 2012)

[6] Burghartz, J.: 'Ultra-thin chip technology and applications' (Springer Science & Business Media, 2010. 2010)

[7] Moonen, P.F., Yakimets, I., and Huskens, J.: 'Fabrication of transistors on flexible substrates: from mass-printing to high-resolution alternative lithography strategies', Advanced materials, 2012, 24, (41), pp. 5526-5541

[8] Khan, S., Lorenzelli, L., and Dahiya, R.S.: 'Technologies for printing sensors and electronics over large flexible substrates: a review', Sensors Journal, IEEE, 2015, 15, (6), pp. 3164-3185

[9] Dahiya, R.S., Adami, A., Collini, C., and Lorenzelli, L.: 'Fabrication of single crystal silicon micro-/nanostructures and transferring them to flexible substrates', Microelectronic Engineering, 2012, 98, pp. 502-507

[10] Zhang, K., Seo, J.-H., Zhou, W., and Ma, Z.: 'Fast flexible electronics using transferrable silicon nanomembranes', Journal of Physics D: Applied Physics, 2012, 45, (14), pp. 143001

[11] Nathan, A., Park, B.-k., Ma, Q., Sazonov, A., and Rowlands, J.A.: 'Amorphous silicon technology for large area digital X-ray and optical imaging', Microelectronics Reliability, 2002, 42, (4), pp. 735-746

[12] Nomura, K., Ohta, H., Takagi, A., Kamiya, T., Hirano, M., and Hosono, H.: 'Room-temperature fabrication of transparent flexible thin-film transistors using amorphous oxide semiconductors', Nature, 2004,

432, (7016), pp. 488-492

[13] Yang, C.-S., Smith, L., Arthur, C., and Parsons, G.N.: 'Stability of low-temperature amorphous silicon thin film transistors formed on glass and transparent plastic substrates', Journal of Vacuum Science & Technology B, 2000, 18, (2), pp. 683-689

[14] Dimitrakopoulos, C.D., and Malenfant, P.R.: 'Organic thin film transistors for large area electronics', Advanced Materials, 2002, 14, (2), pp. 99-117

[15] Sun, Y., and Rogers, J.A.: 'Inorganic semiconductors for flexible electronics', Advanced Materials, 2007, 19, (15), pp. 1897-1916

[16] Xia, Y., Yang, P., Sun, Y., Wu, Y., Mayers, B., Gates, B., Yin, Y., Kim, F., and Yan, H.: 'One-Dimensional Nanostructures: Synthesis, Characterization, and Applications', Advanced Materials, 2003, 15, (5), pp. 353-389

[17] Dasgupta, N.P., Sun, J., Liu, C., Brittman, S., Andrews, S.C., Lim, J., Gao, H., Yan, R., and Yang, P.: '25th anniversary article: semiconductor nanowires–synthesis, characterization, and applications', Advanced Materials, 2014, 26, (14), pp. 2137-2184

[18] Schmidt, V., Wittemann, J.V., and Gösele, U.: 'Growth, Thermodynamics, and Electrical Properties of Silicon Nanowires†', Chemical Reviews, 2010, 110, (1), pp. 361-388

[19] Lu, W., and Lieber, C.M.: 'Nanoelectronics from the bottom up', Nature materials, 2007, 6, (11), pp. 841-850

[20] Cao, G.: ' Nanostructures and Nanomaterials-Synthesis, properties and applications' (World Scientific, 2004. 2004)

[21] Iijima, S.: 'Helical microtubules of graphitic carbon', Nature, 1991, 354, (6348), pp. 56-58

[22] De Volder, M.F., Tawfick, S.H., Baughman, R.H., and Hart, A.J.: 'Carbon nanotubes: present and future commercial applications', Science, 2013, 339, (6119), pp. 535-539

[23] Kim, H.-M., Cho, Y.-H., Lee, H., Kim, S.I., Ryu, S.R., Kim, D.Y., Kang, T.W., and Chung, K.S.: 'High-Brightness Light Emitting Diodes Using Dislocation-Free Indium Gallium Nitride/Gallium Nitride Multiquantum-Well Nanorod Arrays', Nano Letters, 2004, 4, (6), pp. 1059-1062

[24] Mårtensson, T., Svensson, C.P.T., Wacaser, B.A., Larsson, M.W.,

Seifert, W., Deppert, K., Gustafsson, A., Wallenberg, L.R., and Samuelson, L.: 'Epitaxial III–V Nanowires on Silicon', Nano Letters, 2004, 4, (10), pp. 1987-1990

[25] Núñez, C.G., Pau, J., Ruíz, E., Marín, A.G., García, B., Piqueras, J., Shen, G., Wilbert, D., Kim, S., and Kung, P.: 'Enhanced fabrication process of zinc oxide nanowires for optoelectronics', Thin Solid Films, 2014, 555, pp. 42-47

[26] Kind, H., Yan, H., Messer, B., Law, M., and Yang, P.: 'Nanowire Ultraviolet Photodetectors and Optical Switches', Advanced Materials, 2002, 14, (2), pp. 158-160

[27] Núñez, C.G., Marín, A.G., Nanterne, P., Piqueras, J., Kung, P., and Pau, J.: 'Conducting properties of nearly depleted ZnO nanowire UV sensors fabricated by dielectrophoresis', Nanotechnology, 2013, 24, (41), pp. 415702

[28] Núñez, C.G., Braña, A., Pau, J., Ghita, D., García, B., Shen, G., Wilbert, D., Kim, S., and Kung, P.: 'Pure zincblende GaAs nanowires grown by Ga-assisted chemical bedam epitaxy', Journal of Crystal Growth, 2013, 372, pp. 205-212

[29] Núñez, C.G., Braña, A., López, N., and García, B.: 'GaAs nanowires grown by Ga-assisted chemical beam epitaxy: Substrate preparation and growth kinetics', Journal of Crystal Growth, 2015, 430, pp. 108-115

[30] Ross, F.M.: 'Controlling nanowire structures through real time growth studies', Reports on Progress in Physics, 2010, 73, (11), pp. 114501

[31] Björk, M., Ohlsson, B., Thelander, C., Persson, A., Deppert, K., Wallenberg, L., and Samuelson, L.: 'Nanowire resonant tunneling diodes', Applied Physics Letters, 2002, 81, (23), pp. 4458-4460

[32] Thelander, C., Mårtensson, T., Björk, M., Ohlsson, B., Larsson, M., Wallenberg, L., and Samuelson, L.: 'Single-electron transistors in heterostructure nanowires', Applied Physics Letters, 2003, 83, (10), pp. 2052-2054

[33] Wen, C.-Y., Tersoff, J., Hillerich, K., Reuter, M., Park, J., Kodambaka, S., Stach, E., and Ross, F.: 'Periodically changing morphology of the growth interface in Si, Ge, and GaP nanowires', Physical review letters, 2011, 107, (2), pp. 025503

[34] Lu, W., Xiang, J., Timko, B.P., Wu, Y., and Lieber, C.M.: 'One-dimensional hole gas in germanium/silicon nanowire heterostructures', Proceedings of the National Academy of Sciences of the United States of America, 2005, 102, (29), pp. 10046-10051

[35] Fan, H.J., Werner, P., and Zacharias, M.: 'Semiconductor Nanowires: From Self-Organization to Patterned Growth', Small, 2006, 2, (6), pp. 700-717

[36] Wang, N., Cai, Y., and Zhang, R.Q.: 'Growth of nanowires', Materials Science and Engineering: R: Reports, 2008, 60, (1–6), pp. 1-51

[37] Wagner, R., and Ellis, W.: 'Vapor-liquid-solid mechanism of crystal growth and its application to silicon' (Bell Telephone Laboratories, 1965. 1965)

[38] Dick, K.A.: 'A review of nanowire growth promoted by alloys and non-alloying elements with emphasis on Au-assisted III–V nanowires', Progress in Crystal Growth and Characterization of Materials, 2008, 54, (3), pp. 138-173

[39] Duan, X., and Lieber, C.M.: 'General synthesis of compound semiconductor nanowires', Advanced Materials, 2000, 12, (4), pp. 298-302

[40] Zhang, R.Q., Lifshitz, Y., and Lee, S.T.: 'Oxide-Assisted Growth of Semiconducting Nanowires', Advanced Materials, 2003, 15, (7-8), pp. 635-640

[41] Wagner, R.S., Ellis, W.C., Jackson, K.A., and Arnold, S.M.: 'Study of the Filamentary Growth of Silicon Crystals from the Vapor', Journal of Applied Physics, 1964, 35, (10), pp. 2993-3000

[42] Morales, A.M., and Lieber, C.M.: 'A laser ablation method for the synthesis of crystalline semiconductor nanowires', Science, 1998, 279, (5348), pp. 208-211

[43] Wu, Y., Cui, Y., Huynh, L., Barrelet, C.J., Bell, D.C., and Lieber, C.M.: 'Controlled growth a        nd structures of molecular-scale silicon nanowires', Nano Letters, 2004, 4, (3), pp. 433-436

[44] Givargizov, E.: 'Fundamental aspects of VLS growth', Journal of Crystal Growth, 1975, 31, pp. 20-30

[45] Bootsma, G., and Gassen, H.: 'A quantitative study on the growth of silicon whiskers from silane and germanium whiskers from

germane', Journal of Crystal Growth, 1971, 10, (3), pp. 223-234
[46] Kodambaka, S., Tersoff, J., Reuter, M., and Ross, F.: 'Diameter-independent kinetics in the vapor-liquid-solid growth of Si nanowires', Physical review letters, 2006, 96, (9), pp. 096105
[47] Kim, B., Tersoff, J., Kodambaka, S., Reuter, M., Stach, E., and Ross, F.: 'Kinetics of individual nucleation events observed in nanoscale vapor-liquid-solid growth', Science, 2008, 322, (5904), pp. 1070-1073
[48] Weyher, J.: 'Some notes on the growth kinetics and morphology of VLS silicon crystals grown with platinum and gold as liquid-forming agents', Journal of Crystal Growth, 1978, 43, (2), pp. 235-244
[49] Wacaser, B.A., Dick, K.A., Johansson, J., Borgström, M.T., Deppert, K., and Samuelson, L.: 'Preferential interface nucleation: an expansion of the VLS growth mechanism for nanowires', Advanced Materials, 2009, 21, (2), pp. 153-165
[50] Schmidt, V., Senz, S., and Gösele, U.: 'Diameter dependence of the growth velocity of silicon nanowires synthesized via the vapor-liquid-solid mechanism', Physical Review B, 2007, 75, (4), pp. 045335
[51] Kalache, B., i Cabarrocas, P.R., and i Morral, A.F.: 'Observation of incubation times in the nucleation of silicon nanowires obtained by the vapor–liquid–solid method', Japanese journal of applied physics, 2006, 45, (2L), pp. L190
[52] Kikkawa, J., Ohno, Y., and Takeda, S.: 'Growth rate of silicon nanowires', Applied Physics Letters, 2005, 86, (12), pp. 123109
[53] Shakthivel, D., and Raghavan, S.: 'Vapor-liquid-solid growth of Si nanowires: A kinetic analysis', Journal of Applied Physics, 2012, 112, (2), pp. 024317
[54] Dubrovskii, V., Sibirev, N., and Cirlin, G.: 'Kinetic model of the growth of nanodimensional whiskers by the vapor-liquid-crystal mechanism', Technical physics letters, 2004, 30, (8), pp. 682-686
[55] Markov, I.: 'Crystal Growth for Beginners: Fundamentals of Nucleation', Crystal Growth and Epitaxy, 1995, pp. 69
[56] Takei, K., Takahashi, T., Ho, J.C., Ko, H., Gillies, A.G., Leu, P.W., Fearing, R.S., and Javey, A.: 'Nanowire active-matrix circuitry for low-voltage macroscale artificial skin', Nature materials, 2010, 9, (10), pp. 821-826

[57] Ahn, J.-H., Kim, H.-S., Lee, K.J., Jeon, S., Kang, S.J., Sun, Y., Nuzzo, R.G., and Rogers, J.A.: 'Heterogeneous three-dimensional electronics by use of printed semiconductor nanomaterials', science, 2006, 314, (5806), pp. 1754-1757

[58] Freer, E.M., Grachev, O., Duan, X., Martin, S., and Stumbo, D.P.: 'High-yield self-limiting single-nanowire assembly with dielectrophoresis', Nature Nanotechnology, 2010, 5, (7), pp. 525-530

[59] Liu, X., Long, Y.-Z., Liao, L., Duan, X., and Fan, Z.: 'Large-scale integration of semiconductor nanowires for high-performance flexible electronics', ACS nano, 2012, 6, (3), pp. 1888-1900

[60] Ford, A.C., Ho, J.C., Fan, Z., Ergen, O., Altoe, V., Aloni, S., Razavi, H., and Javey, A.: 'Synthesis, contact printing, and device characterization of Ni-catalyzed, crystalline InAs nanowires', Nano Research, 2008, 1, (1), pp. 32-39

[61] Javey, A., Nam, S., Friedman, R.S., Yan, H., and Lieber, C.M.: 'Layer-by-layer assembly of nanowires for three-dimensional, multifunctional electronics', Nano letters, 2007, 7, (3), pp. 773-777

[62] Acharya, S., Panda, A.B., Belman, N., Efrima, S., and Golan, Y.: 'A Semiconductor-Nanowire Assembly of Ultrahigh Junction Density by the Langmuir–Blodgett Technique', Advanced Materials, 2006, 18, (2), pp. 210-213

[63] Panda, A.B., Acharya, S., Efrima, S., and Golan, Y.: 'Synthesis, assembly, and optical properties of shape-and phase-controlled ZnSe nanostructures', Langmuir, 2007, 23, (2), pp. 765-770

[64] Yu, G., Cao, A., and Lieber, C.M.: 'Large-area blown bubble films of aligned nanowires and carbon nanotubes', Nature nanotechnology, 2007, 2, (6), pp. 372-377

[65] Fan, Z., Ho, J.C., Jacobson, Z.A., Yerushalmi, R., Alley, R.L., Razavi, H., and Javey, A.: 'Wafer-scale assembly of highly ordered semiconductor nanowire arrays by contact printing', Nano letters, 2008, 8, (1), pp. 20-25

[66] Fan, D., Zhu, F., Cammarata, R., and Chien, C.: 'Efficiency of assembling of nanowires in suspension by ac electric fields', Applied physics letters, 2006, 89, (22), pp. 223115

[67] Whang, D., Jin, S., Wu, Y., and Lieber, C.M.: 'Large-scale hierarchical organization of nanowire arrays for integrated

nanosystems', Nano letters, 2003, 3, (9), pp. 1255-1259

[68] Tao, A., Kim, F., Hess, C., Goldberger, J., He, R., Sun, Y., Xia, Y., and Yang, P.: 'Langmuir-Blodgett silver nanowire monolayers for molecular sensing using surface-enhanced Raman spectroscopy', Nano letters, 2003, 3, (9), pp. 1229-1233

[69] Wang, D., Chang, Y.-L., Liu, Z., and Dai, H.: 'Oxidation resistant germanium nanowires: Bulk synthesis, long chain alkanethiol functionalization, and Langmuir-Blodgett assembly', Journal of the American Chemical Society, 2005, 127, (33), pp. 11871-11875

[70] Jin, S., Whang, D., McAlpine, M.C., Friedman, R.S., Wu, Y., and Lieber, C.M.: 'Scalable interconnection and integration of nanowire devices without registration', Nano Letters, 2004, 4, (5), pp. 915-919

[71] Marín, A.G., Núñez, C.G., Rodríguez, P., Shen, G., Kim, S., Kung, P., Piqueras, J., and Pau, J.: 'Continuous-flow system and monitoring tools for the dielectrophoretic integration of nanowires in light sensor arrays', Nanotechnology, 2015, 26, (11), pp. 115502

[72] Takahashi, T., Takei, K., Adabi, E., Fan, Z., Niknejad, A.M., and Javey, A.: 'Parallel array InAs nanowire transistors for mechanically bendable, ultrahigh frequency electronics', ACS nano, 2010, 4, (10), pp. 5855-5860

[73] Friedman, R.S., McAlpine, M.C., Ricketts, D.S., Ham, D., and Lieber, C.M.: 'Nanotechnology: High-speed integrated nanowire circuits', Nature, 2005, 434, (7037), pp. 1085-1085

[74] McAlpine, M.C., Friedman, R.S., Jin, S., Lin, K.-h., Wang, W.U., and Lieber, C.M.: 'High-performance nanowire electronics and photonics on glass and plastic substrates', Nano Letters, 2003, 3, (11), pp. 1531-1535

[75] Dahiya, R.S., Metta, G., Valle, M., Adami, A., and Lorenzelli, L.: 'Piezoelectric oxide semiconductor field effect transistor touch sensing devices', Applied Physics Letters, 2009, 95, (3), pp. 034105

[76] Liu, Z., Xu, J., Chen, D., and Shen, G.: 'Flexible electronics based on inorganic nanowires', Chemical Society Reviews, 2015, 44, (1), pp. 161-192

[77] McAlpine, M.C., Ahmad, H., Wang, D., and Heath, J.R.: 'Highly ordered nanowire arrays on plastic substrates for ultrasensitive flexible chemical sensors', Nature materials, 2007, 6, (5), pp. 379-384

[78] Wu, W., Wen, X., and Wang, Z.L.: 'Taxel-addressable matrix of vertical-nanowire piezotronic transistors for active and adaptive tactile imaging', Science, 2013, 340, (6135), pp. 952-957

# CHAPTER FOUR

## Flexible\Stretchable Piezoelectric Nanofiber Devices

Yongqing Duan[1], Yajing Ding[1], Youhua Wang[1], Yewang Su[2], YongAn Huang[1]

[1] *State Key Laboratory of Digital Manufacturing Equipment and Technology, Huazhong University of Science and Technology, Wuhan, 430074, China*

[2] *State Key Laboratory of Nonlinear Mechanics, Institute of Mechanics, Chinese Academy of Sciences, Beijing 100190, China*

## Abstract

Multifunctional capability, flexible design, and low-cost manufacturing are desired attributes for wearable and bio-integrated electronics. Piezoelectric materials, in forms that offer the ability to bend and stretch, are attractive for pressure/force sensors and energy harvesters. This chapter discusses how to fabricate flexible/stretchable piezoelectric devices by the use of electrohydrodynamical direct-writing technique. The electrohydrodynamical direct-writing utilizes stable high electrostatic field and tunable mechanical drawing force to produce piezoelectric PVDF nanofibres with the β-phase formation through in situ electrical poling and mechanical stretching. Bendable piezoelectric devices can be directly fabricated by depositing the poled nanofiber onto a thin polymeric substrate, and generate electricity under bending and lateral pressure. Further, highly stretchable piezoelectric devices are electrohydrodynamically printed by consideration of self-organized in-surface buckling of the fiber-on-substrate system. The electrical output of the stretchable energy harvester is characterized that the generated current is a linear ratio to the stretching frequency and strain, as well as the number of nanofibers. Additionally, the stretchable devices generate electrical energy stably by the lateral pressure even in different stretching status. Experimental and theoretical studies provide detailed insights into the energy conversion mechanisms. This chapter provides detailed engineering design rules and paves a cost-effective and high-efficiency manufacturing pathway for applications in wearable and bio-integrated electronics.

**Keywords:** Electrospinning; Stretchable electronics; Printed electronics; Energy harvester; Piezoelectric materials

## 1. Introduction

An emerging development trajectory in electronics focuses on wearable applications such as health/wellness monitors, surgical tools, smart clothes, and artificial muscles [1], [2]. Likewise, precision tactile sensors might represent first steps toward realization of artificial, electronic skins that mimic the full, multi-modal characteristics and physical properties of natural dermal tissues [3]. Devices that exploit mechanical motions as natural sources of power can be particularly valuable, because of the potential applications in self-powered microelectronics and wireless sensor networks [4-7]. The sum of these small individual energies can be large enough to power electronic systems.

Flexible/stretchable, lightweight piezoelectric devices are key requirements for both sensors and energy harvesters. Conventional piezoelectric materials, which are often in the form of bulk solids or thin films, have low sensitivity in responding to small mechanical forces. They are also too stiff to be fitted with a flexible/stretchable electronics system. One-dimensional nanomaterials have demonstrated an excellent route for scavenging small mechanical energy from ambient environments (such as irregular vibrations [8], airflows [9] and tiny human activities [10-13]). Energy harvesters made of inorganic nanowires have been reported to have higher energy conversion efficiency compared with their micro-sized counterparts [14-17]. However, these inorganic nanowires are brittle and can only work on a very small strain level, which is a big barrier for the integration of inorganic nanowires into wearable electronic devices [18].

Piezoelectric polymers are good alternatives for similar kinds of applications, because they can exploit deformations induced by small force/pressure, mechanical vibration, elongation/compression, bending or twisting [3, 5, 19, 20]. They have higher strain level and thus are able to withstand larger mechanical deformation. Additionally, piezoelectric polymers are attractive for their potential advantages in terms of low manufacturing cost, high resistance to fatigue, and environmental friendliness [21]. Among the known piezoelectric polymers, poly-vinylidene fluoride (PVDF) is the only commercial product used as a piezoelectric membrane, due to its high piezoelectric activity and chemical/mechanical stability [22, 23]. Recently, several papers have reported the use of electrospinning to produce PVDF nanofibers for energy harvesting [3, 24, 25], sensing and actuating [3, 5], either single electrospun PVDF nanofiber [5, 25] or nanofiber mats [18, 26].

Single PVDF nanofiber has been deposited across a pair of electrodes using a near-field electrospinning process to harvest small mechanical vibration [25, 27, 28]. A flexible power generator based on cyclic stretching–releasing of a piezoelectric fine wire was reported to generate an electrical output of 20–50 mV, and 400–750 pA, respectively [12]. However, this type of energy harvester typically faces common problems such as low energy conversion efficiency, complex post-processing, and low stretchability.

Here, we report electrohydrodynamically printed, flexible/stretchable piezoelectric nanofiber devices to meet the requirements of large deformation of bio-integrated electronics and wearable electronics.

Mechanoelectrospinning (MES) process can mechanically stretch and electrically pole the untreated PVDF to transfer non-polarized $\alpha$ and $\gamma$ phases to polarized $\beta$-phase which is necessary for generating piezoelectricity [25]. Additionally, it is able to control the diameter of these fibers. These nanofibers are made of in situ electrically and mechanically poled PVDF with high flexibility, large-deflection sensing, and energy scavenging applications [25]. It is believed that the newly introduced fabrication process could be the manufacturing foundation for a new class of piezoelectric devices made of polymeric nanofibers.

## 2. Experimental Section
### a) Mechanoelectrospinning for Poled PVDF Nanofiber

PVDF exists in several forms, such as $\alpha$ (TGTG'), $\beta$ (TTTT), and $\gamma$ (TTTGTTTG') phases, depending on the chain conformations as trans (T) or gauche (G) linkages. Among them, the $\beta$ phase has the most favorable piezoelectric property which is mainly based on the dipole orientation in the crystalline phase [11]. Thus, the key to get high piezoelectricity of PVDF is to achieve high β crystal phase content and well oriented molecular dipoles. Usually, PVDF can be stretched and poled to β-phase by a strong electric field [14, 22, 25].

Figure 1a shows the schematic of MES system. The nozzle-to-substrate distance ranges from 0.5~10 mm, far smaller than 10~30 cm adopted in traditional electrospinning [29]. Besides a high voltage exerted between nozzle and substrate like traditional electrospinning, MES adopts an additional high-speed motion stage as the role of the person in "Chinese kite" to draw and direct-write nanofibers. The motion stage is digitally controlled and hence stretches the jetting fiber in a programmable manner. Therefore, MES produces fibers based on the uniaxial stretching of a viscoelastic solution by a combination of stable electrostatic field and tunable mechanical force.

Figure 1b illustrates the MES process: filling nozzle with functional ink → adjusting the applied voltage to generate a stable jet from Taylor cone → Moving the substrate to draw the jet. When adjusting the applied voltage, it first gradually increases to a critical value to jet the ink from the apex of the Taylor cone, then decreases to a lower value just to stabilize the Taylor cone. In the last step, the liquid jet undergoes extensive stretching as substrate moving, then attaches onto the substrate orderly. MES is able to precisely manipulate the position, size, and morphology of each electrospun micro/nanofiber. Three key processing parameters are:

1) The nozzle-to-substrate distance, $h_{z\text{-axis}} = h_{straight} + h_{whipping}$, is tunable from 0.5 mm to 10 mm, to regulate the cross-sectional

geometry (i.e. circular or rectangular) as well as the shape (i.e. linear or serpentine) of the written structure.

2) The applied voltage, $U_{voltage}$, which dictates the dynamic behavior of jetting, as well as the fiber diameter.

3) The translational speed, $v_{substrate}$, which controls the mechanical drawing force and hence the fiber diameter or ribbon width. The above three features make MES a versatile direct-writing technique, which can fabricate abundant microstructures such as fiber array [27], serpentine structures [29], and bead-on-string fiber structures [30]. Considering the high electrostatic field and mechanical drawing characteristics, MES is ideally suited for producing piezoelectric nanofibers with the β-phase formation through *in situ* electrical poling and mechanical stretching [27, 31]. MES process can deposit straight PVDF fibers over a large area (Figure 1c). When MES directly writes fibers onto a pre-strained substrate, the fibers will buckle into serpentine structures (Figure 1d) after releasing the pre-strain. MES is a cost-effective and high-efficiency way of direct-writing ultra-thin, flexible/stretchable and ultra-light piezoelectric nanofibers/wires, making them more applicable in scavenging small mechanical energy.

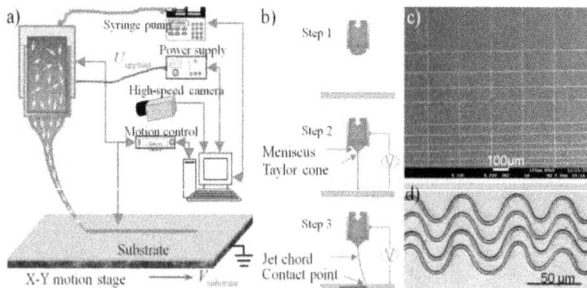

Figure 1: (a) Schematic diagram of mechano-electrospinning process, (b) Steps of MES direct-writing: ink is first filled in the nozzle and a drop forms at the end of the nozzle; when voltage is applied, the end drop is jetted from the nozzle and a Taylor cone is formed; when substrate starts to move, a fine 'jet chord' is formed between the meniscus and the contact point on the substrate. (c) and (d) are the direct-written fibers (straight fibers and in-surface buckled fibers). (Syringe pump: 11 Pico Plus, HARVARD; Powersupply: DW-P403 Dongwen; High-speed camera: Basler A504k)

## b) Fabrication of Bendable PVDF Nanogenerators

PVDF fibers were direct-written onto flexible PET/PI substrate through MES, to obtain high uniform straight fibers for making bendable devices. In the MES process, the PVDF solution [20] was delivered using a syringe pump at a feed rate of 300 nL/min. A stainless steel nozzle was adopted as an electrode, and the ground collector was a metal plate fixed on a moving stage. A high voltage was exerted between nozzle and collector to pull out the jet. The applied voltage was 1.3 - 1.7 kV, the nozzle-to-collector distance was 3 - 10 mm, and the moving speed was 100 - 300 mm/s.

## c) Fabrication of Stretchable PVDF Nanogenerators

PVDF fibers were direct-written onto a prestrained PDMS substrate through MES, to obtain high uniform buckled fibers for making stretchable devices. This approach contains three key steps: prestraining the elastomeric substrate, direct-writing of straight fibers, and releasing the prestrained substrate. In the second step, the PVDF solution was delivered using a syringe pump at a feed rate of 600 nL/min. The applied voltage was 1.5 kV, the nozzle-to-collector distance varied from 4 mm to 10 mm to tune the fiber cross-section shape from ribbon to circle, and the moving speed of the substrate was adjusted from 200 to 400 mm/s. Based on MES, the sectional area and shape (the width and height) of the fiber are tunable via changing process parameters like voltage, speed and nozzle-to-collector distance.

## 3. Results and Discussion
### The Polarization of β Phase in PVDF

The electrohydrodynamically direct-written PVDF nanofibers have high piezoelectricity even without post-processing [5]. Fourier transform infrared (FTIR) spectra of the fibres were collected in attenuated total reflectance mode (ATR) using a VERTEX 70 spectrophotometer (Bruker, Germany). Samples were placed on top of the ATR set and scanned from 1600 to 400 cm−1. The samples are mainly made in three steps: scrape some fibers deposited on Si wafer, add a bit BaCO3 power and pulverize them homogeneously, and press them into bread totally.

Figure 2(a) shows the Fourier transform infrared spectroscopy (FTIR) transmission spectrum of the PVDF fibers electrospun with applied voltages varying from 1.3 kV to 1.7 kV, at the case of V=300mm/s and H=10mm. One can indicate that polar β phase bands appear distinctly at 511, 600, 840 and 1275 cm-1 and the non-polar α phase are not obvious (411, 795, 1402 cm-1) [32, 33]. In order to determine the fraction of the content of β phase, infrared spectroscopy absorption bands at 795 and 840 cm-1 are chosen to characterize the α and β phase. The relative fraction of β phase can be calculated by using [34]:

$$F(\beta) = \frac{X_\beta}{X_\alpha + X_\beta} = \frac{A_\beta}{(K_\beta/K_\alpha)A_\alpha + A_\beta} = \frac{A_\beta}{1.26A_\alpha + A_\beta} \qquad (1)$$

where $K_\alpha = 6.1 \times 10^4 cm^2 / mol$ and $K_\beta = 7.7 \times 10^4 cm^2 / mol$

X is the degree of crystallinity of each phase, A is the absorption bands for $\alpha$ and $\beta$ phase. Figure 2(b) shows that F($\beta$) increases monotonously with the applied voltage, but converges to about 60%. For the electric poling effect, the cells of the $\alpha$ phase chain rotate to align their dipole moments in the direction favored by the external electric field, which increases the $\beta$ phase content. The tunable mechanical drawing force, resulted from moving the substrate, has also large influence on the relatively faction of $\beta$ phase, as shown in Figure 2(c) and (d). When the speed varies from 0 to 350 mm/s, F($\beta$) linearly increases by 8% at the case of U=1.7kV and H=10mm. Larger speed corresponds to larger mechanical drawing force, which results in more chains aligned [27, 31]. It indicates that the mechanical drawing force is effective for $\beta$ phase transformation, and this is an advantage of MES over traditional electrospinning.

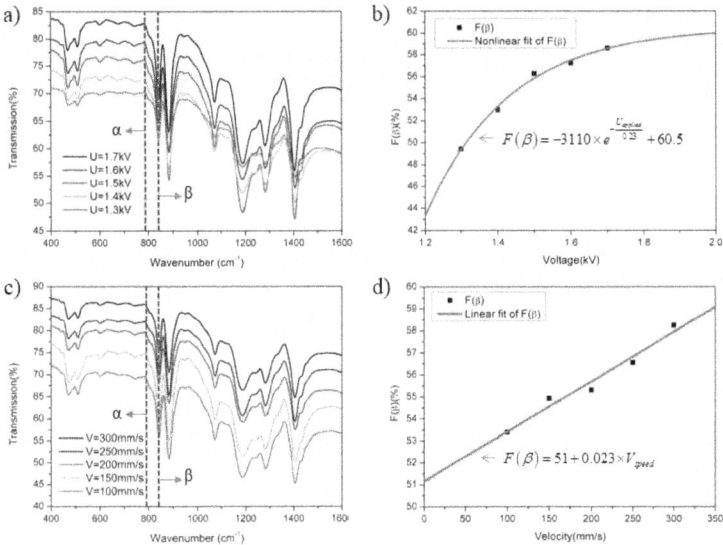

Figure 2: FTIR transmission spectrum of PVDF nanofibers: (a) transmission of PVDF with applied voltage between nozzle and substrate; (b) the relatively faction of the β-phase with applied voltage between nozzle and substrate; (c) transmission of PVDF with the speed; (b) the relatively faction of β-phase with the speed.

## Flexible Piezoelectric Devices Based on Individually Aligned Nanofibers

The flexible piezoelectric device contains four layers as shown in Figure 3a: PET substrate (100 μm), Cu electrode (20 μm), PVDF fibers (~5 μm) and PDMS capsulation layer (~200 μm). The PVDF fibers are not located at the neutral surface since the modulus of PDMS (~2 MPa) is much smaller than those of PET (200 MPa) and Cu (~110GPa). The electrical contacts are established to the ends of fiber arrays on a flexible PET substrate, and the formed devices reveal large current and voltage response to even minute excitation (pressure & bending).

The electric properties of the PVDF fibers were characterized through a semiconductor characterization system (KEITHLEY 4200-SCS) and a probe system (CASCADE SUMMTI 11000). Figure 3b shows the generated current and voltage of a piezoelectric PVDF device under bending. The current and voltage have opposite signs, because the current flows through the PVDF fibers (like current flows through a battery) is from low potential to high potential. The measurements showed a periodic alternation of positive and negative output peaks, corresponding to the application and releasing of the bend, respectively. Both current and voltage responses increase with the PET thickness and the bending frequency and curvature. For example, Figure 3c-d shows the output of a device to cycling bending tests at 1 Hz, 1.6 Hz and 2.3 Hz on a home-made flexural endurance tester. The maximum current (4 nA) and voltage (150 mV) were observed at 2.3 Hz. Up to 1,000 cycles of bending and releasing tests revealed no significant changes in electrical output. These characteristics make fiber arrays promising as building blocks for ultrasensitive piezoelectric devices.

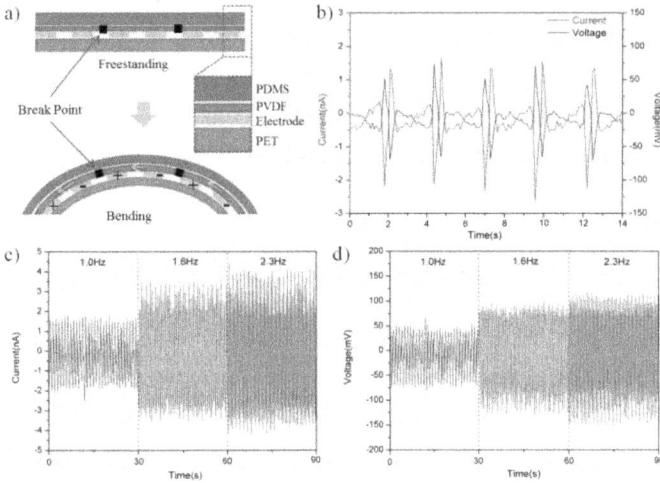

Figure 3: (a) The structure of a bendable PVDF energy harvester; (b) the output current and voltage of the harvester bend and release at a frequency of 0.4 Hz; (c) and (d) are the current and voltage generated at various bending frequencies with 50 PVDF fibers.

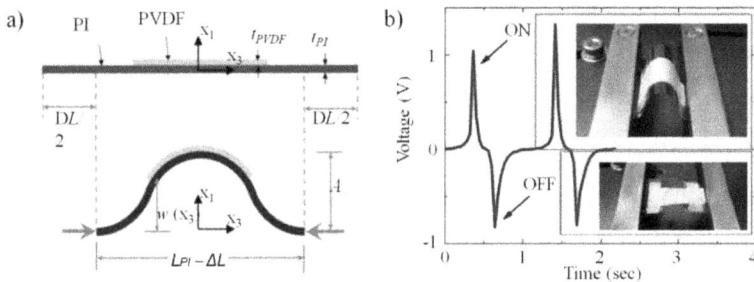

Figure 4: Experimental and theoretical studies of responses of piezoelectric devices. (a) Schematic illustration of an analytical model for the coupling of mechanical deformation and piezoelectric response during bending. (b) Measured voltage response of an array of PVDF fibres under cycling bending at 1 Hz.

During compression, the sample buckled to generate a bent shape (Figure 4a), with curvature consistent with simple mechanics considerations. The measurements showed a periodic alternation of positive and negative output peaks, corresponding to the application and release of the buckling stress, respectively (Figure 4b). The top and bottom insets show photographs of the device during bending and release, respectively. Both responses were increased with PI thickness and bending frequency.

As with applied pressure, simple analytical models can account for the behaviours under bending. Under compression the PI substrate of length LPI buckles into a sinusoidal form represented by the out-of-plane displacement $w = A\left[1 + \cos\left(2\pi x_3/L_{PI}\right)\right]/2$ , where the origin of coordinate $x3$ is at the center of the substrate, and the amplitude A is related to compression $\Delta L$ of the substrate by $A \approx \left(2/\pi\right)\sqrt{L_{PI} \cdot \Delta L}$ [35]. Here the critical compression to trigger buckling, ~2 μm for 225 μm-thick and 6 cm-long PI substrate, is negligible as compared to $\Delta L = 3$ cm in the experiments. The strain along the poling direction x3 at the mid-plane of the fiber array is given by $\varepsilon_{33} = -w''\left(t_{PI} + t_{PVDF}\right)/2$ , where $t_{PI}$ and $t_{PVDF}$ represent the thickness of PI and PVDF. Its strain $\varepsilon_{11}$ along x1 direction and electric field E3 along the poling direction are obtained from the constitutive relation:

$$0 = c_{11}\varepsilon_{11} + c_{13}\varepsilon_{33} - e_{31}E_3 \tag{2}$$

$$D_3 = e_{31}\varepsilon_{11} + e_{33}\varepsilon_{33} + k_{33}E_3 \tag{3}$$

where, D3 is the electric displacement along the poling direction, and it is a constant to be determined. For short-circuit current measurement between two ends of the PVDF fiber array, the voltage across the length of PVDF $L_{PVDF}$ is zero, which, together with the above equations, gives:

$$D_3 = \left[ 2\bar{e}\sqrt{\Delta L/L_{PI}} \left( t_{PI} + t_{PVDF} \right) / L_{PVDF} \right] \sin\left( \pi L_{PVDF}/L_{PI} \right) \tag{4}$$

where $\bar{e} = e_{33} - \left( c_{13}/c_{11} \right)e_{31}$ is the effective piezoelectric constant. The current I is then obtained from $I = -t_{PVDF} w_{PVDF} \dot{D}_3$, where  is the width of PVDF fiber array. For a representative compression $\Delta L = \Delta L_{max} \left[ 1 - \cos\left( 2\pi t/T \right) \right]^2 / 4$ with the maximum compression $\Delta L_{max} = 3$ cm and period T=0.5 and 1 second as in experiments, the maximum current is given by :

$$I_{max} = 2\pi\left( -\bar{e} \right) \frac{\left( t_{PI} + t_{PVDF} \right) t_{PVDF} w_{PVDF}}{L_{PVDF}T} \sqrt{\frac{\Delta L_{max}}{L_{PI}}} \sin\left( \frac{\pi L_{PVDF}}{L_{PI}} \right) \tag{5}$$

For a thickness of PI substrate tPI= 225  as in experiments, Eq. (5) gives the range of Imax=27 nA for T=0.5 s and Imax=14 nA for T=1 s, while experiments give Imax=33 nA and Imax=26 nA for T=0.5 and 1 s, respectively. Here the effective piezoelectric constant for the fiber arrays is taken as $\bar{e} = -2.1\,\mathrm{C/m}^2$, which is larger than that for films (~-0.4 C/m2) [37] because of the strong anisotropy of arrays due to their fibrous structure.

For voltage measurement, V is no longer zero. The electric displacement becomes:

$$D_3 = \left[2\overline{e}\sqrt{\Delta L/L_{PI}}\left(t_{PI}+t_{PVDF}\right)/L_{PVDF}\right]\sin\left(\pi L_{PVDF}/L_{PI}\right)+\left(\overline{k}/L_{PVDF}\right)V \quad (6)$$

The current $I = -t_{PVDF}w_{PVDF}\dot{D}_3$ is also related to the voltage V and the resistance R of the voltmeter by I=V/R, which gives $V/R = -t_{PVDF}w_{PVDF}\dot{D}_3$, or equivalently

$$\frac{dV}{dt}+\frac{L_{PVDF}}{\overline{k}Rt_{PVDF}w_{PVDF}}V = 2\left(-\overline{e}\right)\frac{t_{PI}+t_{PVDF}}{\overline{k}}\sin\left(\frac{\pi L_{PVDF}}{L_{PI}}\right)\frac{d}{dt}\sqrt{\frac{\Delta L}{L_{PI}}}. \quad (7)$$

For $\Delta L = \Delta L_{max}\left[1-\cos\left(2\pi t/T\right)\right]^2/4$ and the initial condition , the maximum voltage is given by:

$$V_{max} \approx 2\pi\left(-\overline{e}\right)\frac{R}{T}\frac{\left(t_{PI}+t_{PVDF}\right)t_{PVDF}w_{PVDF}}{L_{PVDF}}\sin\left(\frac{\pi L_{PVDF}}{L_{PI}}\right)\sqrt{\frac{\Delta L_{max}}{L_{PI}}} \quad (8)$$

For a thicknesses of PI substrate tPI=225   and a resistance of the voltmeter R = 70 M MΩ in experiment, Eq. (8) gives a range of Vmax=2.1 V for T=0.5 s and Vmax=0.85 V for T=1 s, while experiments give Vmax=1.3 V and Vmax=1.0 V for T=0.5 and 1 s, respectively [3].

## Stretchable Energy Harvesters Based on In-Surface Buckled Nanofibers

The stretchable piezoelectric devices can be fabricated by direct-writing the straight nanofibers on a pre-strained substrate using MES process, followed by releasing the pre-strain. The buckling modes of nanofibers, such as out-of-surface buckling and in-surface buckling as shown in Figure 5(a, b), are determined by the cross-section shape of the nanofibers [5], which can be tuned by process parameters (e.g. nozzle-to-substrate distance, applied voltage, moving speed of substrate). The electrospun nanofibers were examined with a laser scanning confocal microscopy (LSCM, KEYENCE VK-X200K). The fiber diameter and cross-section were obtained from LSCM images via image analysis software VK-X Series. It can be noted from Figure 5 (c, d) that the applied voltage and the nozzle-to-substrate distance have a different effect on the cross-section morphology. The former controls the flow of solution at the nozzle, and larger flow corresponds to larger width and height of the deposited microstructures. The latter controls the solidification degree of the jetting fiber, and higher solidification results in a thinner and taller cross-sections. Figure 5d shows that the extreme is circular cross-section with equal width and height.

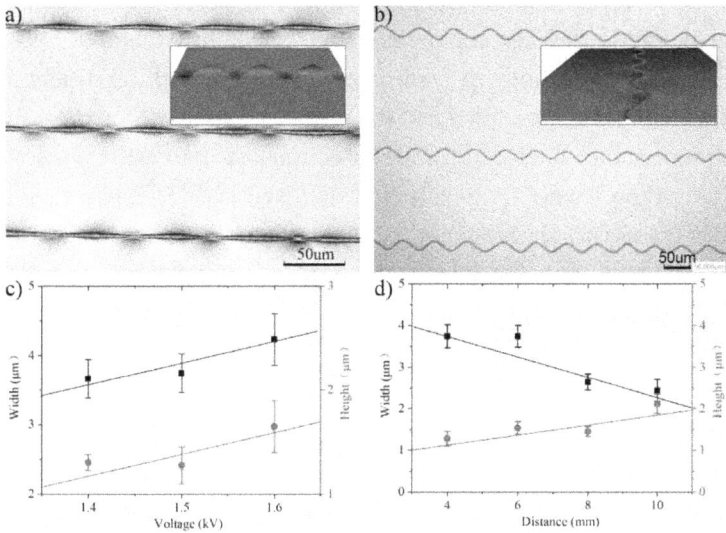

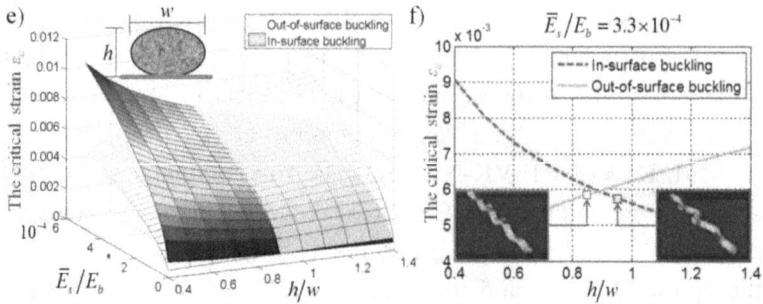

Figure 5: (a) Out-of-surface buckled fibers; (b) In-surface buckled fibers); (c) The width and height of electrospun fibers versus the applied voltage; (d) The width & height of electrospun fibers versus the nozzle-to-substrate distance; (e) The critical strain versus the height-to-width ratio of fiber and the Young's modulus ratio of substrate to fiber; (f) The critical strains for two buckling modes at E⁻_S/E_b=3.3× 〚10 〛 ^(-4)

As in-surface buckled fibers are more suitable for stretchable piezoelectric devices for their stable mechanical behavior, then the question is how to generate in-surface buckled fibers by adjusting MES parameters, or what's the key factors that influence the buckling mode/direction. A mechanical model through the least-energy principle is discussed [36]. For a given fiber-on-substrate system, the buckling mode with lower total energy (including the bending energy and membrane energy of fibers, and substrate energy) is favorable.

The deflection of the micro/nanofiber on the substrate can be described as $v_1 = v_{m1} \cos k_1 x$ and $v_2 = v_{m2} \cos k_2 x$ (Subscript 1 and 2 denote out-of-surface buckling and in-surface buckling respectively), where $v_{m1}$ and $v_{m2}$ are the buckling amplitude, $k_1 = 2\pi / \lambda_1$, $k_2 = 2\pi / \lambda_2$, $\lambda_1$ and $\lambda_2$ are the buckling wavelength, and the coordinate x is along the fiber axis.

For the out-of-surface buckling system, the bending energy per unit length of the buckled micro/nanofiber is:

$$U_{b1} = \frac{k_1}{2\pi} \int_0^{\frac{2\pi}{k_1}} \frac{1}{2} E_b I_{b1} (\frac{\partial^2 v_1}{\partial x^2})^2 dx = \frac{E_b I_{b1}}{4} k_1^4 v_{m1}^2 \quad (9)$$

where $E_b$ is Young's modulus of fiber and $I_{b1}$ is the moment of inertia in out-of-surface direction.

The membrane strain of fiber is $\varepsilon_{m1} = du_1 / dx + (dv_1 / dx)^2 / 2$, $u_1$ and $v_1$ represent axial displacement and normal deflection respectively. Since Young's modulus of PVDF fiber is much larger than that of PDMS substrate, the effect of shear stress on buckling is negligible, which gives a constant membrane strain in the buckled fiber as $\varepsilon_{m1} = k_1^2 v_{m1}^2 / 4 - \varepsilon_{pre}$, where $-\varepsilon_{pre}$ is the compressive strain in the micro/nanofiber due to the relaxation of the stretched substrate. The membrane energy in the micro/nanofiber is:

$$U_{m1} = \frac{1}{2} E_b A \varepsilon_{m1}^2 = \frac{1}{2} E_b A (\frac{1}{4} v_{m1}^2 k_1^2 - \varepsilon_{pre})^2 \quad (10)$$

where   is the area of cross-section of the fiber.

The elastomeric substrate is modelled as a semi-infinite solid, since its thickness is several orders of magnitude larger than that of the micro/nanofiber. The out-of-surface buckled fiber imposes a normal stress traction to the surface of the substrate, which takes the form of $T_1 = P_1 \cos(k_1 x)$, where $P_1 = -E_b A v_{m1} k_1^2 (k_1^2 v_{m1}^2 / 4 - \varepsilon_{pre}) - E_b I_{b1} v_{m1} k_1^4$ [37]. Due to this traction, the normal displacement of a point on the top surface of distance   y   to   the   micro/nanofiber   center   is   [38]

$$v_{sub1} = P_1 \cos(k_1 x)[W(1-\gamma+\ln 2)-(\frac{W}{2}+y)\ln(k_1|\frac{W}{2}+y|)-(\frac{W}{2}-y)\ln(k_1|\frac{W}{2}-y|)]/(\pi\bar{E}_s\frac{W}{2})$$

where,  $\bar{E}_s = E_s/(1-v_s^2)$ is  the  plane  strain  modulus  of
substrate,  $E_s$  and  $v_s$  are the elastic modulus and Poisson's ratio
of the PDMS substrate,  $\gamma = 0.577$  is Euler's constant and W  is
the width of the interfacial contact region. The strain energy per
unit length in the substrate can be obtained via divergence
theorem                                                                    as:

$$U_{s1} = \frac{k_1}{2\pi}\int_0^{\frac{2\pi}{k_1}}\int_{-W/2}^{W/2}\frac{P_1\cos(k_1 x)}{2W}v_{sub1}dydx = \frac{P_1^2}{4\pi\bar{E}_s}(3-2\gamma-2\ln\frac{k_1 W}{2}) \quad (11)$$

The total energy in the micro/nanofiber-on-substrate can be
obtained as:

$$U_{tot1} = U_{b1} + U_{m1} + U_{s1} - \frac{1}{\lambda_1}\int_0^{\lambda_1} P_1 \cos k_1 x(v_{sub1} - v_{m1}\cos k_1 x)dx \quad (12)$$

$$= -\frac{E_b I_{b1}}{4}k_1^4 v_{m1}^2 + \frac{1}{2}E_b A(\varepsilon_{pre} - \frac{1}{4}v_{m1}^2 k_1^2)(\varepsilon_{pre} + \frac{3}{4}v_{m1}^2 k_1^2) - \frac{P_1^2}{4\pi\bar{E}_s}(3-2\gamma-2\ln\frac{k_1 W}{2})$$

where, the last integration represents the work across the
fiber/substrate interface39. The minimization of  with respect to
and  gives:

$$U_{tot1} = \frac{1}{2}E_b A\varepsilon_{c1}(2\varepsilon_{pre} - \varepsilon_{c1}) \quad (13)$$

$$\varepsilon_{c1} \approx \sqrt{\frac{\bar{E}_S}{E_b}}\frac{\sqrt{I_{b1}}}{A}(\frac{9}{16}+\frac{16\pi}{9}/(3-2\gamma-2\ln\frac{3}{4}-\frac{1}{2}\ln\frac{W^4}{16}\frac{\bar{E}_S}{E_b I_{b1}})) \quad (14)$$

where  $\varepsilon_{c1}$  is the critical buckling strain, or the minimum strain
needed to induce buckling. The micro/nanofiber only compresses
when  $\varepsilon_{pre} \le \varepsilon_{c1}$, and buckles out-of-surface when  $\varepsilon_{pre} > \varepsilon_{c1}$.

For the in-surface buckling system, the bending energy $U_{b2}$ and membrane energy $U_{m2}$ of the micro/nanowire are the same calculation with out-of-surface buckling, i.e. $U_{b2} = \frac{E_b I_{b2}}{4} k_2^4 v_{m2}^2$,

$U_{m2} = \frac{1}{2} E_b A (\frac{1}{4} v_{m2}^2 k_2^2 - \varepsilon_{pre})^2$, $I_{b2}$ is the moment of inertia in in-surface direction. The substrate energy $U_{s2}$ is different as the substrate bear lateral force in in-surface buckling while normal force in out-of-surface buckling. The lateral stress traction also takes the form of $P_2 \cos(k_2 x)$, where

$P_2 = -E_b A v_{m2} k_2^2 (k_2^2 v_{m2}^2 / 4 - \varepsilon_{pre}) - E_b I_{b2} v_{m2} k_2^4$ [37]. Due to the lateral stress traction, the lateral displacement of a point on the top surface of distance y to the micro/nanowire center is

$v_{sub2} = P_2 \cos(k_2 x)[W(\frac{1}{1-\upsilon_s} - \gamma + \ln 2) - (\frac{W}{2} + y)\ln(k_2 | \frac{W}{2} + y |) - (\frac{W}{2} - y)\ln(k_2 | \frac{W}{2} - y |)]/(\pi \bar{E}_s \frac{W}{2})$

[40]. The strain energy of substrate per unit length is:

$U_{s2} = \frac{k_2}{2\pi} \int_V \sigma_{ij} \varepsilon_{ij} dV = \frac{P_2^2}{4\pi \bar{E}_s} (\frac{3-\upsilon_s}{1-\upsilon_s} - 2\gamma - 2\ln \frac{k_2 W}{2})$ . The minimum

energy of the in-surface buckling system can be represented as:

$U_{tot2} = U_{b2} + U_{m2} + U_{s2} - \frac{1}{\lambda_2} \int_0^{\lambda_2} P_2 \cos k_2 x (v_{sub2} - v_{m2} \cos k_2 x) dx$ \hfill (15)

$= -\frac{E_b I_{b2}}{4} k_2^4 v_{m2}^2 + \frac{1}{2} E_b A(\varepsilon_{pre} - \frac{1}{4} v_{m2}^2 k_2^2)(\varepsilon_{pre} + \frac{3}{4} v_{m2}^2 k_2^2) - \frac{P_2^2}{4\pi \bar{E}_s} (\frac{3-\upsilon_s}{1-\upsilon_s} - 2\gamma - 2\ln \frac{k W}{2})$

where the last integration represents the work across the fiber/substrate interface [39]. The minimization of $U_{tot2}$ with respect to $v_{m2}$ and $k_2$ gives:

$U_{tot2} = \frac{1}{2} E_b A \varepsilon_{c2} (2\varepsilon_{pre} - \varepsilon_{c2})$ \hfill (16)

$$\varepsilon_{c2} \approx \sqrt{\frac{\bar{E}_S}{E_b}} \frac{\sqrt{I_{b2}}}{A} (\frac{1}{2} + 2\pi / (5 - 2\gamma - 2\ln\frac{5}{7} - \frac{1}{2}\ln\frac{W^4}{16}\frac{\bar{E}_S}{E_b I_{b2}})) \qquad (17)$$

where $\varepsilon_{c2}$ is the critical buckling strain or the minimum strain for in-surface buckling. The micro/nanofiber buckles in-surface once the pre-strain $\varepsilon_{pre}$ reaches $\varepsilon_{c2}$.

For a micro/nanofiber on an elastomeric PDMS substrate system, the buckling mode with smaller total energy is favorable. Comparison of the two total energies through Eqs. 13 and 16 gives the conclusion that the buckling mode with smaller critical buckling strain is energetically favorable.

To clarify the influence of fiber cross-section to buckling mode, the critical buckling strains for elliptical cross-sections [5], which are close to the actual fiber sections, are studied as shown in Figure 5e. When $E^-S/E\_b = 3.3 \times [\![10]\!] ^\wedge(-4)$ ($E\_b = 2$ "GPa" for PVDF fiber, and $E\_s = 0.5$ "MPa" for PDMS substrate), buckling mode transformation point (intersection point) is $h/w = 0.88$ as shown in Figure 5f, i.e. in-surface buckling is always favorable if . The green (red) rectangles in Figure 5f represent finite element simulation results of out-of-surface (in-surface) buckling with different aspect ratios, which is consistent with the theoretical analysis. In-surface buckled fibers can be kept in the surface during the fiber-on-substrate system pressed or stretched which will be discussed later. It establishes the foundation for the stable generation of electrical output under different stretching conditions.

Stretchable piezoelectric devices made up of in-surface buckled fibers are fabricated and tested as shown in Figure 6a. The stretchable energy harvester is fixed on a home-made tensile platform. When the tensile platform reciprocates, the piezoelectric PVDF nanofibers generate current or voltage which is characterized by a semiconductor characterization system (KEITHLEY 4200-SCS).

Figure 6b shows the current (~1.2 nA) and voltage (~40 mV) measured from a stretchable piezoelectric device (3 cm long between electrodes) consisting of ~120 in-surface buckled PVDF fibers under cyclic tensile and release (30% strain, 0.5 Hz) test. The results demonstrate both high stretchability and high piezoelectricity of the fibers.

Figure 6c shows that the output current increases with the applied strain   (20 fibers, 0.5Hz). The applied strains  are 40%, 70% and 100%, and the output current increases from 0.33 nA to 0.75 nA and 1.1 nA.

Figure 6d exhibits the average maximum currents are 0.65 nA, 1.62 nA and 2.8 nA, corresponding to the frequencies   of 0.2 Hz, 0.5 Hz, 0.8 Hz, respectively (40 fibers, $\varepsilon_{applied} = 70\%$). It can be found from Figure 6 (e) and (f) that the output current increases linearly with the stretching frequency and the number of the fibers, respectively. These results show that the performance characteristics of buckled piezoelectric PVDF fibers are consistent with fundamental piezoelectric theory $i_{fiber} = d_{33} E_{fiber} A_{fiber} \dot{\varepsilon}_{applied}$ [19], where $i_{fiber}$ is the generated current, $d_{33}$ is the piezoelectric charge constant, $E_{fiber}$ is the Young's modulus, the cross-sectional area $A_{fiber}$ is proportional to the number of fibers, and the applied strain rate $\dot{\varepsilon}_{applied}$ is proportional to $\varepsilon_{applied} \times f_{applied}$. When more PVDF fibers are integrated, and stretched with larger strain and frequency, the response current is larger. Since MES is able to manipulate fibers individually, a stretchable piezoelectric generator with accurate output current can be fabricated.

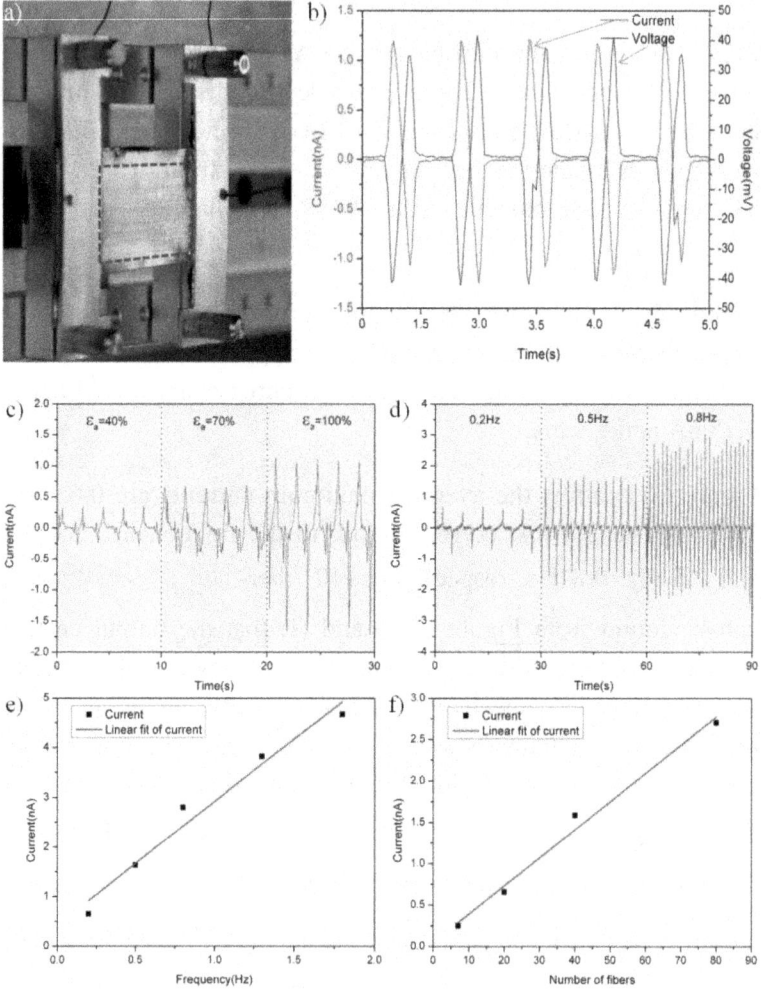

Figure 6:  Response of stretchable piezoelectric devices. (a) Photograph of the tensile test platform. (b) Output current and voltage of device consisting of 120 PVDF fibers measured with respect to time under an applied strain of 30% at 0.5 Hz. (c) and (d) are the piezoelectric behavior of in-surface buckled fibers on PDMS substrate tested with different applied strains and different cycling frequencies. (e) The output current versus the stretch and release frequency (70%, 40 fibers) (f) The output current versus the number of fibers (70%, 0.5 Hz).

Figure 7(a) and (b) show that when the device is stretched under different applied strains, its sensitivity to pressure is maintained ideally. The electrical respondence of in-surface buckled fibers to pressure is not affected by the external tensile strain obviously. The main reason is that the buckled fibers can keep in surface at various stretching status, as shown in

Figure 7(c) and (d). The in-surface buckled fibers avoid the sudden change of deformation of fibers when subjected to the vertical compression. This characteristic makes in-surface buckled fibers very suitable for applications of stretchable pressure sensors under various tensions. It should be emphasized that buckled PVDF fibers on elastomeric substrates can be stretched over 100% applied strain (limited by the failure of PDMS), much larger than the strain (~8%) of the reported piezoelectric generator or sensor [41].

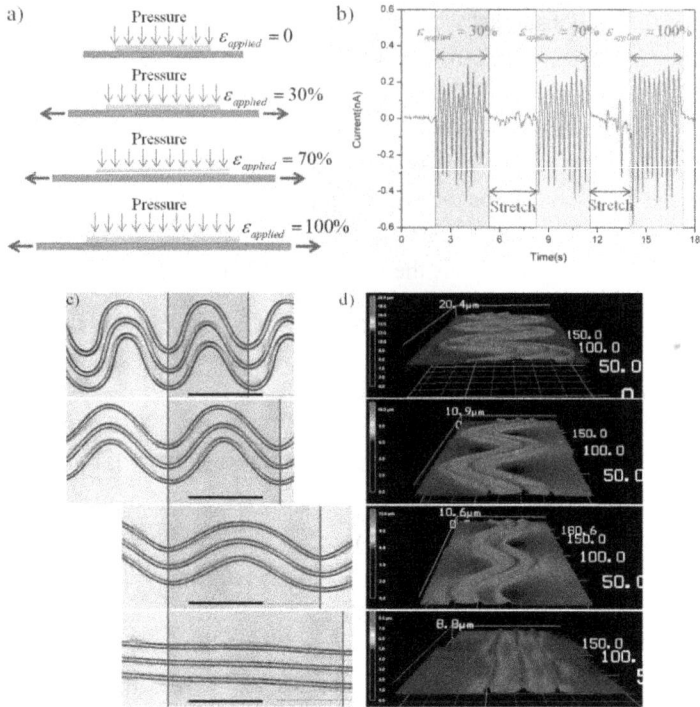

Figure 7: The piezoelectric effect of in-surface buckled PVDF device under stretching status: (a) schematic diagram of piezoelectric device under various stretching status; (b) the current generated under 30%, 70% and 100% applied strains; (c) and (d) are the top view and the oblique view of in-surface buckled fibers under the applied strain 0%, 30%, 70%, and 100%. The bars denote 50 μm.

To evaluate the sensitivity under pressure quantitatively, a soft elastomer delivered well defined levels of pressure to the arrays [3]. Devices, formed simply by establishing electrical contacts to the ends of a ribbon shaped sample of fibre arrays on a flexible support, reveal large response to even minute applied pressures. The aligned PVDF fibre arrays were placed on 75–150 and 225 μm thick kapton film, and electric connections were established with copper films (25 μm thick) and silver paint (Ted Pella Fast Dring Silver Paint, 160040-30). Open loop voltage measurements were performed by using a DAQ (SMU2055) USB multimeter (6.5 digit resolution, Agilent Technologies) with an input resistance of R = 70 MΩ. Short-circuits current measurements were performed with a Semiconductor Parameter Analyzer (4155C Agilent Technologies).

Figure 8b shows well-behaved, linear variations in the output voltage with pressure, for various values of the effective contact areas between 9 and 36 mm2 (squares with sides Leff;

Figure 8a), with slopes between 0.41 and 0.79 mV/Pa. For a given pressure (10 Pa, Leff = 3 mm), the output voltage does not change significantly with length of the fiber array.

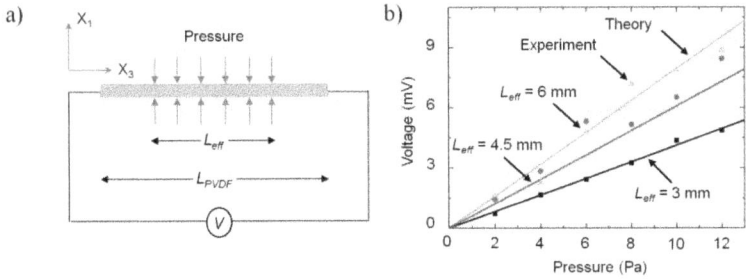

Figure 8: The output responses of piezoelectric devices. (a) Schematic illustration of an analytical model for the response of arrays of PVDF fibres under applied compression -p along x1 direction over the effective contact length (Leff). LPVDF is the total length of the PVDF fibre array. (b) Experimental (symbols) and theoretical (lines) pressure response curves at different Leff.

Theoretical model was established to discover the mechanism of electricity generation [3]. The analytical models can capture the piezoelectric behaviors. Consider a resultant projected component of piezoactive dipoles along the longitudinal axis of the PVDF nanofibers, x3, as induced by the deformation and external force (Figure 8a). The fiber array is isotropic with elastic, piezoelectric, and dielectric constants cij, eij, and kij, respectively. For an applied compression -p along the x1 direction over an effective contact length Leff, the strain   and electric field E3 along the poling direction are obtained from the constitutive relation

$$-p = c_{11}\varepsilon_{11} - e_{31}E_3 \qquad (18)$$

$$D_3 = e_{31}\varepsilon_{11} + k_{33}E_3 \qquad (19)$$

where, D3 is the electric displacement along the poling direction, and is zero for vanishing free charge density on the surface; the in-plane strains $\varepsilon_{22} = \varepsilon_{33} \approx 0$ because the PVDF fiber arrays (elastic modulus ~ 200 MPa) are bonded to the much thicker and stiffer plastic substrate (elastic modulus 2.5 GPa). The voltage across Leff is then given by:

$$V = \frac{\overline{d}}{\overline{k}} L_{eff} p \qquad (20)$$

where, $\overline{d} = e_{31}/c_{11}$ and $\overline{k} = k_{33} + e_{31}^2/c_{11}$ are the effective piezoelectric and dielectric constants, respectively. For $\overline{d}/\overline{k} = 0.14 \text{ V} \cdot \text{m/N}$, Eq. (20) agrees well with the experimental results shown in Figure 8a for a wide range of pressures p and the three effective contact lengths Leff = 3, 4.5 and 6 mm used in the experiments [3]. The value of $\overline{d}/\overline{k}$

reaches or exceeds those achieved in films with extreme stretching and poling (e.g., ~ 0.045-0.094 Vm/N 42). Eq. (20) also suggests that the voltage is independent of the total length LPVDF of the PVDF fiber arrays, which is also consistent with experimental results of Figure 8b.

## 4. Conclusion

Highly bendable and stretchable nanogenerators with individually controlled fibers are directly written onto PET substrate and pre-strained elastomeric PDMS substrate by using MES, respectively. The process realizes direct-write the PVDF fiber and in situ obtain piezoactive -phase in one step, without further electrical poling. Stretchable piezoelectric generator exhibit excellent response in extremely large applied strain (more than 100%). Additionally, a stretchable piezoelectric device with in-surface buckled fiber array shows the stable current output under various applied strains owing to its in-surface deformation. The theoretical models for bending and stretching are established to show the mechanism of electrical generation. The collective results suggest that stretchable piezoelectric PVDF devices can be fabricated by cost-effective electrohydrodynamic direct-writing, with attractive stretchability and potential to be integrated into stretchable electronics.

### Acknowledgment

The authors acknowledge supports from the National Natural Science Foundation of China (51322507, 91323303, 51421062) and New Century Excellent Talents in University (NCET-11-0171). The general characterization facilities were provided through the Wuhan National Laboratory for Optoelectronics.

# References

[1] Kim, D.H. et al. Epidermal electronics. Science 333, 838-843 (2011).

[2] Hwang, S.-W. et al. A physically transient form of silicon electronics. Science 337, 1640-1644 (2012).

[3] Persano, L. et al. High performance piezoelectric devices based on aligned arrays of nanofibers of poly (vinylidenefluoride-co-trifluoroethylene). Nat. Commun. 4, 1633 (2013).

[4] Qi, Y. et al. Piezoelectric ribbons printed onto rubber for flexible energy conversion. Nano Lett. 10, 524-528 (2010).

[5] Duan, Y., Huang, Y., Yin, Z., Bu, N. & Dong, W. Non-wrinkled, highly stretchable piezoelectric devices by electrohydrodynamic direct-writing. Nanoscale 6, 3289-95 (2014).

[6] Wang, Z.L. & Song, J. Piezoelectric nanogenerators based on zinc oxide nanowire arrays. Science 312, 242-246 (2006).

[7] Lee, M. et al. Self-powered environmental sensor system driven by nanogenerators. Energy & Environmental Science 4, 3359-3363 (2011).

[8] Hsu, C.-L. & Chen, K.-C. Improving piezoelectric nanogenerator comprises ZnO nanowires by bending the flexible PET substrate at low vibration frequency. The Journal of Physical Chemistry C 116, 9351-9355 (2012).

[9] Sun, C.L., Shi, J., Bayerl, D.J. & Wang, X.D. PVDF microbelts for harvesting energy from respiration. Energy & Environmental Science 4, 4508-4512 (2011).

[10] Xu, S. et al. Self-powered nanowire devices. Nature nanotechnology 5, 366-373 (2010).

[11] Calvert, P. Piezoelectric polyvinylidene fluoride. Nature 256, 694 (1975).

[12] Hansen, B.J., Liu, Y., Yang, R.S. & Wang, Z.L. Hybrid Nanogenerator for Concurrently Harvesting Biomechanical and Biochemical Energy. Acs Nano 4, 3647-3652 (2010).

[13] Lee, M. et al. A hybrid piezoelectric structure for wearable nanogenerators. Adv. Mater. 24, 1759-1764 (2012).

[14] Yu, H. et al. Enhanced power output of an electrospun PVDF/MWCNTs-based nanogenerator by tuning its conductivity. Nanotechnology 24(2013).

[15] Zhu, G.A., Yang, R.S., Wang, S.H. & Wang, Z.L. Flexible High-Output Nanogenerator Based on Lateral ZnO Nanowire Array. Nano Lett. 10, 3151-3155 (2010).

[16] Chen, X., Xu, S.Y., Yao, N. & Shi, Y. 1.6 V Nanogenerator for

Mechanical Energy Harvesting Using PZT Nanofibers. Nano Lett. 10, 2133-2137 (2010).

[17] Chang, J.Y., Domnner, M., Chang, C. & Lin, L.W. Piezoelectric nanofibers for energy scavenging applications. Nano Energy 1, 356-371 (2012).

[18] Fang, J., Niu, H.T., Wang, H.X., Wang, X.G. & Lin, T. Enhanced mechanical energy harvesting using needleless electrospun poly(vinylidene fluoride) nanofibre webs. Energy & Environmental Science 6, 2196-2202 (2013).

[19] Sirohi, J. & Chopra, I. Fundamental understanding of piezoelectric strain sensors. Journal of Intelligent Material Systems and Structures 11, 246-257 (2000).

[20] Ding, Y., Duan, Y. & Huang, Y. Electrohydrodynamically Printed, Flexible Energy Harvester Using In Situ Poled Piezoelectric Nanofibers. Energy Technology (2015).

[21] Fang, X.Q., Liu, J.X. & Gupta, V. Fundamental formulations and recent achievements in piezoelectric nano-structures: a review. Nanoscale 5, 1716-1726 (2013).

[22] Lovinger, A.J. Ferroelectric polymers. Science 220, 1115-1121 (1983).

[23] Holmes-Siedle, A., Wilson, P. & Verrall, A. PVdF: An electronically-active polymer for industry. Mater. Design 4, 910-918 (1984).

[24] Liu, Z.H., Pan, C.T., Lin, L.W., Huang, J.C. & Ou, Z.Y. Direct-write PVDF nonwoven fiber fabric energy harvesters via the hollow cylindrical near-field electrospinning process. Smart Materials and Structures 23(2014).

[25] Chang, C.E., Tran, V.H., Wang, J.B., Fuh, Y.K. & Lin, L.W. Direct-Write Piezoelectric Polymeric Nanogenerator with High Energy Conversion Efficiency. Nano Lett. 10, 726-731 (2010).

[26] Li, D. & Xia, Y. Electrospinning of nanofibers: Reinventing the wheel? Adv. Mater. 16, 1151-1170 (2004).

[27] Bu, N., Huang, Y. & Yin, Z. Continuously tunable and oriented nanofiber direct-written by mechano-electrospinning. Mater. Manuf. Process. 27, 1318-1323 (2012).

[28] Bu, N., Huang, Y., Duan, Y. & Yin, Z. Process Optimization of Mechano-Electrospinning by Response Surface Methodology. J. Nanosci. Nanotechnol. 13, 1-9 (2013).

[29] Huang, Y.A. et al. Versatile, kinetically controlled, high precision electrohydrodynamic writing of micro/nanofibers. Sci. Rep. 4, 5949 (2014).

[30] Bu, N.B., Huang, Y.A., Deng, H.X. & Yin, Z.P. Tunable bead-on-string microstructures fabricated by mechano-electrospinning. J. Phys. D. Appl. Phys. 45, 405301 (2012).

[31] Huang, Y., Wang, X., Duan, Y., Bu, N. & Yin, Z. Controllable self-organization of colloid microarrays based on finite length effects of electrospun ribbons. Soft Matter 8, 8302-8311 (2012).

[32] Sencadas, V., Moreira, V.M., Lanceros-Mendez, S., Pouzada, A.S. & Gregorio, R. alpha-to-beta transformation on PVDF films obtained by uniaxial stretch. Advanced Materials Forum Iii, Pts 1 and 2 514-516, 872-876 (2006).

[33] Andrew, J.S. & Clarke, D.R. Effect of electrospinning on the ferroelectric phase content of polyvinylidene difluoride fibers. Langmuir 24, 670-672 (2008).

[34] Ramos, M.M.D., Correia, H.M.G. & Lanceros-Mendez, S. Atomistic modelling of processes involved in poling of PVDF. Comp. Mater. Sci. 33, 230-236 (2005).

[35] Song, J. et al. Mechanics of noncoplanar mesh design for stretchable electronic circuits. J. Appl. Phys. 105(2009).

[36] Duan, Y.Q., Huang, Y.A. & Yin, Z.P. Competing buckling of micro/nanowires on compliant substrates. J. Phys. D. Appl. Phys. 48(2015).

[37] Timoshenko, S.P. Strength of materials, (Van Nostrand, New York, 1956).

[38] Xiao, J. et al. Mechanics of buckled carbon nanotubes on elastomeric substrates. J. Appl. Phys. 104, 033543 (2008).

[39] Jiang, H.Q. et al. Finite width effect of thin-films buckling on compliant substrate: Experimental and theoretical studies. J. Mech. Phys. Solids 56, 2585-2598 (2008).

[40] Xiao, J. et al. Mechanics of nanowire/nanotube in-surface buckling on elastomeric substrates. Nanotechnology 21, 085708 (2010).

[41] Qi, Y. et al. Enhanced piezoelectricity and stretchability in energy harvesting devices fabricated from buckled PZT ribbons. Nano Lett. 11, 1331-1336 (2011).

[42] Omote, K., Ohigashi, H. & Koga, K. Temperature dependence of elastic, dielectric, and piezoelectric properties of "single crystalline" films of vinylidene fluoride trifluoroethylene copolymer. J. Appl. Phys. 81, 2760-2769 (1997).

# CHAPTER FIVE

# Flexible Silicon Nanoelectronics

Galo Torres Sevilla and Muhammad Mustafa Hussain

*Integrated Nanotechnology Lab, Integrated Disruptive Electronic Applications (IDEA) Lab*

*Computer Electrical Mathematical Science and Engineering Division, 4700 King Abdullah University of Science and Technology (KAUST), Thuwal 23955-6900, Saudi Arabia*

## 1. Introduction

Silicon-based devices dominate today's state-of-the-art nanoelectronics. Since the introduction of the first monolithic microprocessor in the early 70s, nanoelectronics have become an essential part of our digital world. Silicon processing appears to be a suitable candidate for use in flexible, high-performance devices that satisfy emerging market demands in wearable and implantable electronics. In this chapter, we explore chief approaches towards the development of flexible silicon electronics. We begin with Section 2, which describes some of the most promising methodologies for converting rigid silicon substrates into flexible electronics through the use of silicon-based nanomaterials, etch-release processes, and stress-based wafer spalling. In Section 3, we present a thorough study of the most advanced transistor architecture (FinFET) onto a flexible silicon platform, and we report on two different release methods. No significant variations in the key parameters of the transistor were evident by comparing flexible and bulk devices. Finally, we describe the relationship between temperature and flexible FinFETs in subsection 3.6.

## 2. State-Of-The-Art Flexible Silicon Electronics

### 2.1 Transfer Printing of Silicon Nanoribbons and Nanomembranes [1-13]

Single-crystal silicon wafers are some of the most advanced materials being used in conventional consumer electronics and optoelectronics. Advantages such as a high level of purity and a favorable cost/yield make silicon semiconductors popular for use in the most advanced electronic components. In addition, careful control of doping concentrations and a high surface smoothness make monocrystalline silicon an ideal candidate for high-performance thin-film transistors. After being released from standard silicon wafers, nanometer-scale ribbons and membranes can be used as basic building blocks for complex, flexible, high-performance circuitry. During the release process, nanomaterials have the advantage of inheriting the high-quality properties of their source wafers that include advantageous electrical characteristics they can impart to the overall device. In 2004, the first method of releasing nanoribbons (NRs) and nanomembranes (NMs) from monocrystalline silicon wafers was reported [2]. Since then, significant advances in device performance, transfer yield, and the flexibility of fabricated devices and circuits have been made [3-8].

NRs and NMs are obtained by starting with high-quality single-crystal silicon wafers (111) or silicon-on-insulator (SOI) wafers. Next, photolithography is used to define lateral NR and NM dimensions, while their thickness is determined by processing the wafers with chemical etching techniques. On Si wafers (111), silicon anisotropic etching is performed using potassium hydroxide (KOH) or tetramethyl ammonium hydroxide (TMAH). Si (111) wafers can easily be etched along the (110) plane while keeping the (111) plane intact (i.e., wafers are etched along their surfaces at much higher etch rates than depths). Meanwhile, removing NRs and NMs from SOI wafers requires that the wafers be processed with anisotropic and isotropic etching. The first anisotropic etch step removes the unwanted silicon layer, exposing the bottom silicon dioxide ($SiO_2$) and defining the lateral dimensions of the membranes or ribbons. The second isotropic etch step removes the bottom $SiO_2$ layer using hydrofluoric acid (HF), releasing NRs and NMs from the source wafer in preparation for transfer to the final substrate. NRs and NMs are released from the source substrate by adhering to a soft stamp made of an elastomer such as poly-dimethylsiloxane (PDMS) that is placed with some pressure on top of them and transferred to the destination substrate. Adhesion between silicon NRs and the elastomer is dominated by van der Waals interactions. After the NRs and NMs are anchored to the PDMS stamp, they are printed on the surface of the target substrate with microscale precision. Printing may be done using different mechanisms that range from viscoelastic effects to biomimetic strategies to the use of bonding layers [9, 10]. The use of high-resolution cameras during transfer and printing enables submicron-scale control of positioning and registration of the nanomaterials. Using this process, transfer printing of NRs and NMs can be automated to yield transfers of millions of nanomaterials per hour over large or predefined areas, as is required by today's high-performance systems [1]. The extremely thin nature of nanomaterials invokes up to 15 orders of magnitude less mechanical stiffness compared to their source substrates [11]. Since NRs and NMs are usually transferred to very thin, polymer-based substrates, they can be used to fabricate high-performance electronics and sensors capable of small bending radii of curvature (< 1mm) [12]. The first step to fabricating flexible electronic devices involves designing the building block of today's digital world: a metal-oxide-semiconductor field-effect transistor (MOSFET). To fabricate n- and p-type MOSFETs, doped NRs and NMs can be used as a channel material. After the required doped nanomaterials are transferred to the target substrate using the process described above, field-effect transistors are fabricated using state-of-the-art deposition and etching techniques. First, the gate dielectric and gate metal are deposited on top of the printed NRs and NMs. Next, the gate is patterned using standard photolithography and chemical etching techniques. Finally, the source and drain contacts are deposited and defined using the same procedure used for the gate. After device fabrication, the interlayer dielectric (ILD) can be deposited or spin coated and the substrate can be processed again to obtain multiple interconnected layers of devices. There are many important features about this fabrication process. First, low-temperature processes prevent thermal expansion of materials and damage in multilayer systems caused by unwanted deformations. Second, low-temperature substrates and interlayer polymeric materials ensure that the underlying systems are not degraded during multilayer fabrication. Third, the use of soft elastomers as the stamp material enables transfer printing of NRs and NMs to planar or topographic surfaces without affecting the yield (> 99%) of the transferred nanomaterials [1]. Figure 2.1.1 [13] shows the process flow to fabricate flexible silicon NR- and NM-based devices using transfer-printing techniques.

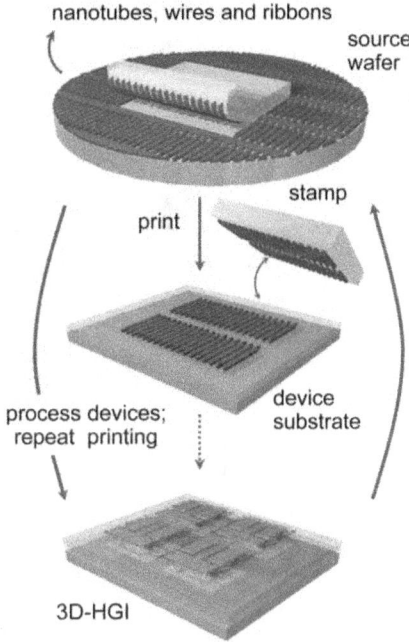

Figure 2.1.1: From [13]. Reprinted with permission from AAAS. Schematic illustration of a printed semiconductor nanomaterials–based approach to heterogeneous 3D electronics.

Fabricated transistors exhibit electrical behavior similar to their rigid counterparts. Figure 2.1.2 [13] shows the electrical characteristics of fabricated MOSFETs using transfer printing of single-crystal silicon NRs and NMs. Curves illustrate extracted mobilities as high as 500 cm$^2$ V$^{-1}$ s$^{-1}$ in the linear region and 100 cm$^2$ V$^{-1}$ s$^{-1}$ in the saturation region; fabricated transistors achieved $I_{on}/I_{off}$ ratios of $10^3$ [5].

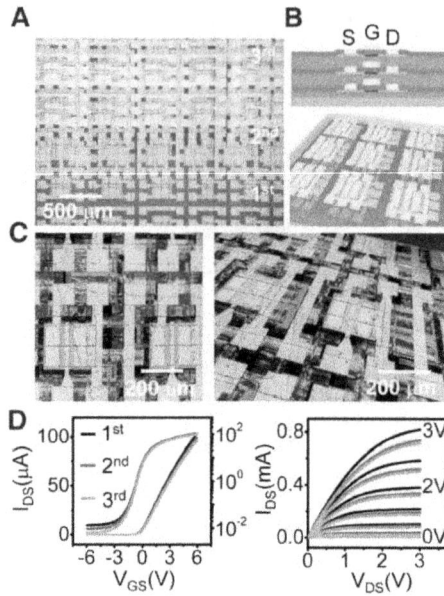

Figure 2.1.2: From [13]. Reprinted with permission from AAAS. (A) Optical micrograph view of the top of a 3D multilayer stack of arrays of single-crystal silicon MOSFETs that use printed silicon nanoribbons for the semiconductor. (B) Schematic cross-sectional (top) and angled (bottom) views. (C) 3D images collected by confocal microscopy on a device substrate similar to that shown in (A) and (B). (D) Current-voltage characteristics of Si MOSFETs in each of the layers, showing excellent performance (mobilities of 470 ± 30 cm2/Vs) and good uniformity in the properties. The channel lengths and widths are 19 and 200 μm, respectively. The overlap lengths, as defined by the distance that the gate electrode extends over the doped source and drain regions, are 5.5 μm. IDS, drain current; VGS, bias voltage; VDS, drain voltage.

The use of 3D n-MOS-based inverters fabricated on two different levels provided the circuit-level characteristics. Drive transistors (*L= 4μm and W=200μm*) consist of an array of NRs placed on the first layer, while the load transistor (*L=4μm and W=30μm*) consists of a second NR array located on a second device layer. Layers are separated using a polymer-based ILD and are connected using e-beam evaporated metal deposited on lithographically defined openings. Figure 2.1.3 [13] shows the device characteristics for the fabricated n-MOS inverter. With a supply voltage of 5 V, the double-layer inverter shows competitive behavior and well-defined transfer characteristics with gains of ~2, comparable to performances typically reported for conventional planar, rigid inverters that use similar transistors. Arranging inverters in series or in parallel configurations can provide different functions, such as ring oscillator capacity and multilayer transfer printing, which can sophisticate devices such that the circuit elements are interconnected through the interlayer dielectric vias, providing 3D integration of silicon NR- and NM-based circuits [13].

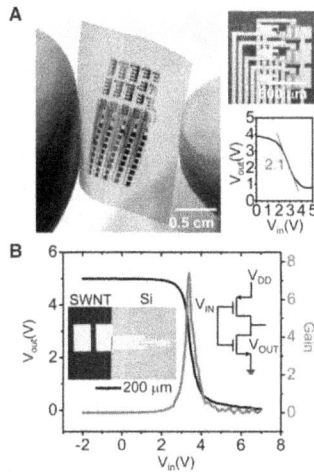

Figure 2.1.3: From [13]. Reprinted with permission from AAAS. (A) Image of a printed array of 3D silicon n-channel metal oxide semiconductor inverters on a PI substrate. (B) Transfer characteristics of a printed complementary inverter that uses a p-channel SWNT TFT (channel length and width of 30 and 200 μm, respectively) and an n-channel Si MOSFET (channel length and width of 75 and 50 μm, respectively).

Figure 2.1.4 shows the high-frequency characteristics of fabricated devices [5]. The cutoff frequency was obtained in the range of 515 MHz for a drain bias of 2 V and a gate bias of 2 V, indicating that reducing channel length can improve the high-frequency operation of the transistor. This high frequency operation shows competitive behavior while approaching the performance levels required to support high-frequency systems such as active antennas.

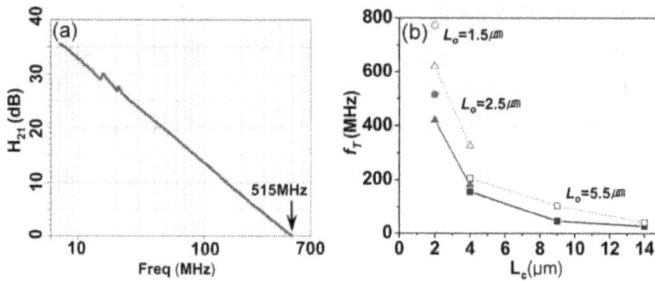

Figure 2.1.4: © [2006] IEEE. Reprinted, with permission, from [5]. (a) Experimental plots of $H_{21}$ obtained from TFT of $L_c = 2$ μm and $L_o = 1.5$ μm with $V_g = 2$ V and $V_d = 2$ V. (b) Dependence of fT on the gate length of TFTs: Measured (filled) and calculated values (open).

Mechanical flexibility is a key feature for the application of fabricated devices to implantable and wearable electronics. NR- and NM-based MOSFET flexibility can be evaluated using custom-made mechanical stages that bend the host polymer-based substrate to generate concave (compressive strain) and convex (tensile strains) curvatures [5]. As expected, devices show minimum performance degradation in terms of carrier mobility; Figure 2.1.5 shows the change in normalized mobility when comparing flat with bent states. Furthermore, reliability studies confirm the robustness of the devices. After 2000 bending cycles with a minimum bending radius of 3 mm, key parameters of the transistors remain less than 20% changed, as shown in Figure 2.1.5B. Since stress-related device degradation is directly dependent on the vertical position of NRs and NMs and the thickness of the host substrate, the lifetime of the devices under bending and unbending conditions may be increased by reducing the thickness of the host substrate or by coating the top of the substrate with thin polymers to locate active devices in neutral planes [5].

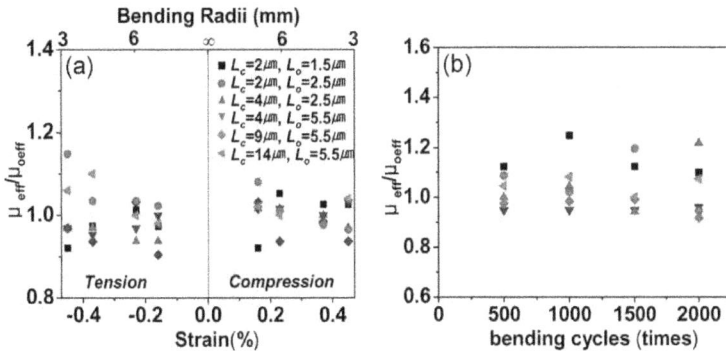

Figure 2.1.5: © [2006] IEEE. Reprinted, with permission, from [5]. (a) Normalized effective mobility ($\mu_{eff}$ /$\mu_{0eff}$ ) as a function of bending-induced strain and bending radius. (b) Normalized effective mobility after bending (to 3-mm radius; 0.44% strain) and unbending the devices several thousand times.

In this section, we have explained that silicon-based nanomaterials in the form of very thin monocrystalline NRs and NMs have high mechanical flexibility and can be used to fabricate transistors with high mobilities and excellent electrical characteristics [1-13]. The advantage of low-temperature processing, allows these devices to be fabricated on polymer-based substrate without damaging the system due to unwanted deformations. Furthermore, fabricated devices show good behavior in the high-frequency region of operation, hence confirming that these NR- and NM-based devices may be used in demanding applications such as radio frequency communications. In addition, inherited polymer properties, such as transparency, can be applied to create new application possibilities. Future work may focus on further development of transfer-printing materials and methods for integrating fabricated devices into real life applications. Also, issues related to electrical and mechanical characteristics such as the semiconductor/dielectric interface and integration with ultramodern fabrication methods are still problems that need to be addressed before transfer printing can become a widely used process in the fabrication of high-performance devices for consumer electronics. A heterointegration with organic materials, such as polymer semiconductors and inorganic carbon nanotube-based films, may enable the fabrication of more robust and stable systems. A key attribute of the transfer printing approach is that it exploits well-developed and standard silicon materials. It also reaches yields that may support applications for wearable and implantable electronics through well-based circuit design and fabrication technologies. Also, the reliability of interconnections, interlayer materials, and adhesion between the active materials and the host substrate may enable the integration of complex systems on flexible polymer-based materials. A few challenges to the mechanical integrity of these fabricated devices need to be resolved before this methodology will be appropriate for use in the design of high-performance electronics. Despite the few drawbacks described, numerous advantages suggest that with further investigation, a more sophisticated technology will ensue for use in a wide range of flexible electronic applications.

## 2.2 Chip-film Technology [14, 15]

One of the critical prerequisites for silicon nanofabrication is the production of sufficiently thick wafers that are reliably mechanically stable during the device-batch fabrication process. Silicon inherits highly beneficial mechanical properties in the form of stiffness and mechanical reliability, but by the end of the fabrication process, wafers have also normally thinned down to enable thermal dissipation and chip-packaging capabilities. Although these thinned wafers are usually fragile and prone to fracture [16], and thus need to be handled delicately to prevent damage to the fabricated circuits and systems, they are also highly desirable for emerging applications in wearable and implantable electronics [17,18]. Back grinding, one of the most common methods for producing thin wafers and chips[19], requires that complementary metal-oxide-semiconductor (CMOS) nanofabrication processes are performed on wafers with starting thicknesses ranging from 500 to 800 µm, depending on the size of the substrate. Once all processes have been completed, the wafers are thinned down to 75 to 300 µm before being diced into single chips, making them susceptible to stress-related fractures that could be induced by rough surfaces or edges; thinner wafers require extra efforts to relieve stress and surface planarization [20]. Controlling silicon thickness is, therefore, a challenge to back-grinding techniques. Fortunately, SOI wafers have improved thickness control by the use of a buried oxide (BOX) in the etch stop layer [21]. SOI wafers can be thinned down by creating trenches on their edges, exposing the BOX for removal by an isotropic etchant such as hydrofluoric acid. On the downside, the high aspect ratio required to etch the trenches is impractical, and handling costs can increase the final cost of ownership of fabricated devices. For this reason, Chip-film™ technology introduced a new approach for creating ultra-thin chips that eliminates the constraints of silicon back grinding and SOI-based thinning.

Figure 2.2.1 [15] illustrates how thin monocrystalline silicon chips are produced using Chip-film technology. This methodology takes an additive approach for defining the final thickness of the chip rather than the subtractive nature of conventional back-grinding techniques. One disadvantage of this technique is that it is a two-step process: pre- and post-processes, which increases the final cost of the fabricated chip. For this reason, Chip-film consists of processing steps during the early stage of fabrication that reduces the risk of incurring additional costs due to yield loss [14].

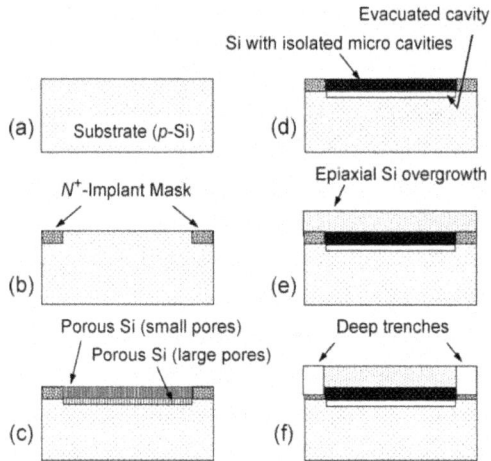

Figure 2.2.1: © 2009, IEEE. Reprinted, with permission, from [15]. Schematic illustration of the Chip-film process flow. (a) Starting p-type substrate. (b) With n+-implant mask. (c) After dual-porous silicon etching within chip areas, resulting in a 1-μm fine-porous layer over a 200-nm coarse-porous layer. (d) After sintering in hydrogen, resulting in a ~1-μmthick microcavity-rich silicon layer and a 200-nm buried continuous cavity. (e) After epitaxial overgrowth with device-quality silicon (~ 20 μm). (f) After trench etching at the chip edges down into the buried cavity, leaving only silicon anchors in selected places.

The Chip-film process begins with single-sided polished wafers with a resistivity of 14 mΩ.cm [15]. Next, N-type regions are formed with ion-implantation of phosphorus and an annealing step to protect the anchor regions from the following anodic etching steps, creating a dual porous silicon layer in the phosphorous undoped regions [22]. The resulting layers consist of a 1-μm-thick layer with 20-30% fine porosity located on top of a second 200-nm-thick layer with 50-60% coarse porosity. Next, a 30-min annealing step at 1100 °C in hydrogen transforms these porous silicon layers with large surface areas into a material comprising monocrystalline silicon with embedded micro/nanocavities and a minimum integral surface area. As a result, the fine porous layer converts into a cavity-rich monocrystalline silicon layer, while the coarse porous layer transforms into a layer with narrow and continuous cavities. In Chip-film technology, the small bottom-buried cavities merge with larger ones to form continuous buried cavities [23]. Meanwhile, the surface of the top layer forms a nearly cavity-free topography that serves as a high-quality seed layer for epitaxial silicon growth. Epitaxial growth is performed at 1100 °C in SiHCl₃ at 760 Torr. The thickness of the epitaxial layer defines the thickness of the final fabricated chip and provides a high-quality surface for CMOS integration [15]. Chip-film wafers can be used as standard wafers for CMOS integration with the minor difference that the chip areas, surrounded by the anchoring regions, have depressions at approximately 250 nm. These depressions can be explained by contributions from the 200-nm-thick buried continuous cavity layer and the 50-nm material loss of the 1-μm-thick porous layer [14]. To integrate the nm-scaled CMOS, a registration with a 250-nm shift in the chip areas with respect to the alignment marks is required for correct device fabrication. After the CMOS integration flow has been completed, trenches are etched along the chip suspending them in a continuous cavity layer exposed at the bottom. Because the chips are now weakly attached to the substrate in the anchoring region, they can be picked up with a high vacuum tool from the top surface of the chip and moved to the final host substrate or package. This process is called Pick, Crack, and Place [14,15]. Note that chips must be supported by the high vacuum tool during the entire transfer and packaging process, otherwise residual tensile or compressive stress from the fine porous silicon layer and the layers deposited during CMOS fabrication cause them to wrap around on themselves. Figure 2.2.2 [14] shows a picture of the final fabricated, released chips using Chip-film technology.

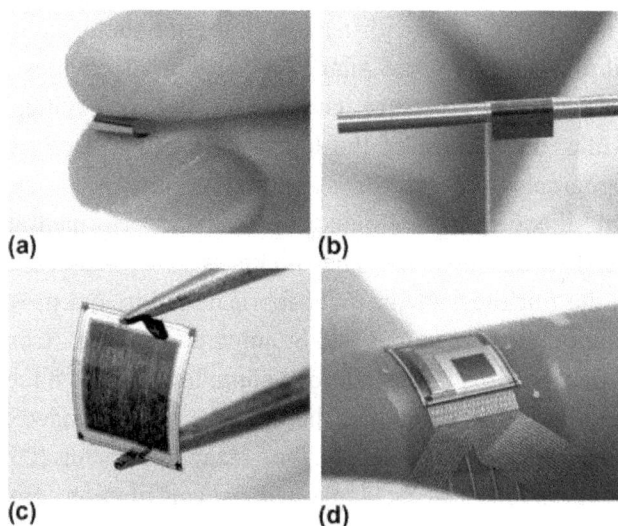

Figure 2.2.2. Reprinted from [14]. Copyright (2010), with permission from Elsevier. Chip-film™ samples (20 μm) under tensile stress for illustration purposes: (a) between fingers, (b) forced into radial bending over a 2-mm metal cylinder, (c) with a mixed-signal integrated circuits between tweezers, and (d) with an RF ID tag glue-attached to a foils substrate between fingers

To test the reliability and viability of Chip-film™ technology to produce thin, flexible silicon chips with high-performance electronics, test devices were fabricated using standard 6'' 0.8-μm CMOS technology. Transistors were fabricated based on a p-well concept that comprises one poly- and two metal layers. To compare their performance, standard- and Chip-film-processed wafers were fabricated simultaneously. Although alignment marks are located in the anchoring regions on the Chip-film wafers, these 200-nm depressions do not impede the fabrication process. Previous work has identified that more advanced CMOS gate technologies require a customized lithography process that prevents misalignments between the first and subsequent layers that can occur as a result of a shift in the focus image between the alignment marks and the chip regions [14,15]. Fabricated transistor characteristics can be found in table 2.2.I [15]. Minor differences in carrier mobilities may be explained by the use of different substrates during the fabrication of bulk and Chip-film devices ($p^+$-Si, 14 mΩ.cm with a 6-μm-thick n-type epitaxial layer for Chip-film and $n^+$-Si, 25-50 mΩ.cm with a 6-μm-thick n-type epitaxial later for bulk devices). Slightly higher mobilities observed with the Chip-film™ technology fabrication process confirm that the surface of the defect-free epitaxial silicon layer is CMOS compatible. A mixed-signal circuit was fabricated containing 38000 digital and 2700 analog transistors to demonstrate the viability of Chip-film™ technology [14]. The results were very similar when comparing Chip-film™ wafers with bulk wafers, hence confirming the defect-free nature of the epitaxial layer grown for Chip-film chip fabrication. Also, to test the chips under bending conditions, Chip-film chips are attached to a Kapton sheet containing cooper interconnections, wire bonded, and tested under 110 MPa stress conditions at a bending radius of 20 mm. Performance degradation measured within the ranges specified for similar commercial chips. Finally, standby current dependence with stress conditions was found to be directly proportional to the bending radius of the chip. However, results show that dependence of the supply current on bending radii can be effectively controlled by changing the physical orientation of the NMOS and PMOS transistors.

Table 2.2.I. © 2009, IEEE. Reprinted, with permission, from [15]. Statistical CMOS transistor and ring oscillator parameters.

|  | Chipfilm™ | Bulk Control | Difference |
|---|---|---|---|
| $\mu_n$ (cm$^2$/Vs) (9x0.8 μm$^2$) | Mean = 633 σ = 10.7 | Mean = 607 σ = 9.4 | +4.3% |
| $\mu_p$ (cm$^2$/Vs) (12x0.9 μm$^2$) | Mean = 178 σ = 4.03 | Mean = 174 σ = 2.1 | +2.3% |
| $L_{eff,n}$ (μm) (0.8 μm) | Mean = 0.79 σ = 0.018 | Mean = 0.8 σ = 0.097 | -1.25% |
| $L_{eff,p}$ (μm) (0.9 μm) | Mean = 0.82 σ = 0.021 | Mean = 0.825 σ = 0.1 | -0.6% |
| $V_{th,n}$ (V) (9x0.8 μm2) | Mean = 0.87 σ = 0.0058 | Mean = 0.87 σ = 0.005 | 0 |
| $-V_{th,p}$ (V) (12x0.9 μm2) | Mean = 0.917 σ = 0.0044 | Mean = 0.915 σ = 0.0044 | 0 |
| $T_{RO}$ (ps) | Mean = 310 σ = 6 | Mean = 300 σ = 6 | +3.3% |

In summary, Chip-film technology complements currently established silicon chip fabrication and eliminates constraints inherent to back grinding and SOI-based release. Improved control of chip thickness is achieved from the extremely controllable nature of epitaxial growth using Chip-film technology. For these reasons, Chip-film is a suitable candidate for the fabrication of ultra-thin chips and for application in wearable and implantable high-performance electronics. Furthermore, the reliability and repeatability of this new technology lends itself toward 3D integration, and finally, the low-cost fabrication of ultra-thin chips using post processes that are commonly used in today's chip fabrication and assembly flows favor the integration of Chip-film technology [14, 15].

## 2.3 Spalling Process [24 – 26]

In the late 80s, researchers discovered that if a tensile stress layer is deposited on the surface of a brittle substrate, the layer will peel away from the surface, taking with it a thin portion of the substrate [27, 28]: this process is called spalling. An analytical model for this process was developed in the late 1980s and offers a direct means of predicting the critical stress loading conditions and crack depth in the substrate. This fracture mode results from the load created by the tensile layer on the edges of the substrate giving rise to normal and shear stresses, which guide the crack in the in the substrate to an equilibrium state below the surface [29]. Until now, most research in spalling has aimed to eliminate material failures and cracking substrates. However, in this Section, we describe how spalling could be used as a means of fabricating thin monocrystalline silicon chips and flexible electronics. In most cases, spalling is conducted by first depositing thick metallic layers, annealing them at temperatures as high as 900 °C, and then cooling the metal. The residual stress caused by thermal expansion mismatches creates spontaneous fractures in the substrate and leads to exfoliation of the top surface [30]. Although layer thickness is controllable and reliable using this method, the thermal budget required for this process makes it incompatible with standard micro-sized and nanofabrication techniques, hence hindering the usefulness of the spalling technique. For this reason, and to make spalling compatible with industry standards, four basic elements need to be addressed. First, the metal stressor layer needs to be deposited with intrinsic stress to avoid high-temperature processing. Second, the stressor layer must not initiate spontaneous spalling, for this reason, reliable crack control must be developed. Third, mechanical control over subsurface cracks is needed to prevent parasitic fracture modes in the surface of the substrate. And fourth, a methodology for handling spalled films needs to be developed. Figure 2.3.1 [24] shows the process flow for controlled spalling that satisfies the requirements mentioned above for the release of the monocrystalline silicon thin layers from bulk "mother" wafers. The process requires applying the stressor layer to and then a handling layer for the released thin silicon film, initiating a crack near the edge of the substrate, and finally mechanically guiding the fracture across the substrate. More details of this process are provided throughout this section.

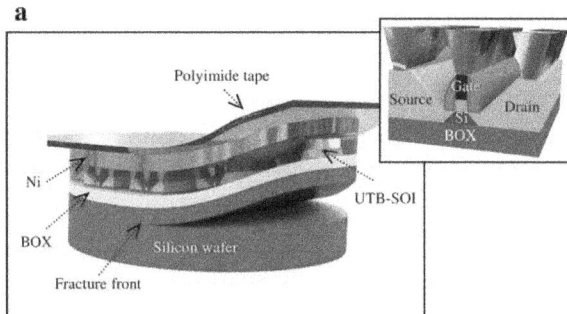

Figure 2.3.1: Reprinted with permission from (25). Copyright (2012) American Chemical Society. Schematic illustration of the controlled spalling process used for removing the prefabricated devices and circuits from the rigid silicon handle wafer. The inset schematically shows the device architecture for the UTB transistors with raised source/drain regions.

A wide range of materials can be used as a thick metallic stressor layer. However, in terms of cost and ease of deposition, nickel (Ni) has proven to be the best candidate. During Ni deposition, film stress can easily be controlled in DC sputtering systems by controlling the partial Ar pressure in the deposition chamber [31], and the electrochemical deposition of nickel provides higher control over intrinsic film stress and reduces operation costs [32]. To test the viability of this process, the Ni-stressor layer was deposited while controlling stress such that it is sufficiently tensile on the edges of the substrate and low enough that it will cause spontaneous spalling on the surface of the wafer. Results show that 175-μm-thick layers could be released from a single Ge (001) monocrystalline wafer. Note, however, that if the thickness of the Ni stressor layer is below the subcritical region shown in Figure 2.3.2, spalling will initiate, but if the Ni-stressor layer is above the optimal thickness, spontaneous spalling of the film could be induced with no control over the spalled films. Therefore, optimal thickness is essential for obtaining desired results; it can be obtained from Suo and Hutchinson's model [28]. Figure 2.3.2b [24] shows the relationship between spalled and Ni-stressor layer thickness. At any one time, a direct relationship between the uniformity of the Ni layer and the spalled film is evident. After the stressor layer is deposited and the intrinsic stress is confirmed to be in the optimal region for spalling, the process to initiate the crack in the subsurface region of the "mother" wafer takes place. Although there are a number of procedures that can be followed to initiate a crack in the monocrystalline wafer, such as laser cutting, the simplest method consists on abruptly terminating the Ni film near the edge of the wafer because a discontinuous stressor film creates a large stress gradient along the free edge of the substrate [33]. The opposing nature of the stress near the edges of the wafer can cause peeling and sheer stress in the GPa range within the substrate. When the handle layer is applied over the stressor layer, and a small force is applied near the edge of the substrate, a crack will form in the subsurface of the wafer that can propagate mechanically with the aid of the handle layer. A film that will not crack in undesired regions requires a handling layer that will control the propagation of single subsurface cracks. Because wafers are subjected to high stress during and after stressor-layer deposition, they need to be held flat by vacuum chucks during the entire spalling process. For example, to successfully remove the handling layer, a 25-$\mu$m-thick polyimide film was applied on top of the stressor layer. Once the handling layer is positioned on the wafer, a small force normal to the surface of the substrate is applied, and the crack is manually directed until the top layer of the wafer is released.

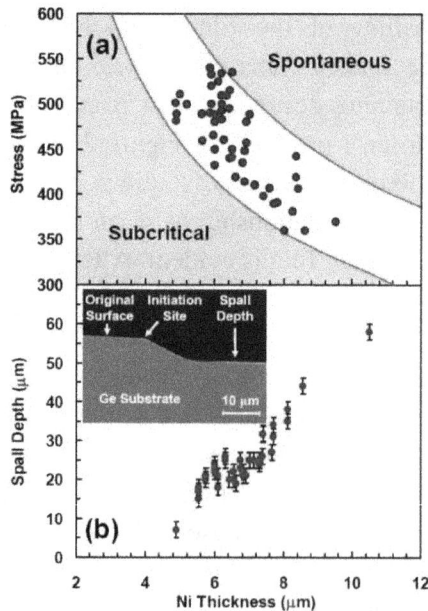

Figure 2.3.2: Reprinted with permission from [24]. (a) Process window for controlled spalling of Ge 100 substrates using sputter deposited Ni layers. The data points represent combinations of Ni stress and thickness that do not lead to spontaneous (self-initiated) spalling, but permit controlled spalling once the handle layer is applied. (b) Measured relationship between Ni thickness and depth of spalling in Ge 100. Depth control of spalling is achieved by adjusting the thickness of the deposited Ni and the intrinsic tensile stress to remain within the process window shown in (a).

The spalling process is capable of constructing large-scale thin monocrystalline silicon layers that can be used to produce flexible high-performance electronics. Although defect-free, and high-efficiency solar cells have been demonstrated using spalling [34], their application to high-performance CMOS-based circuitry has yet to be demonstrated. To show the versatility of the spalling process and its use in CMOS technology, complete circuitry was obtained from standard monocrystalline SOI wafers. Figure 2.3.3 [24] shows how complete CMOS circuits can be removed from their "mother" wafer using spalling technology. By applying an Ni-stressor layer to a 300-mm wafer, circuits 100 mm in diameter are removed selectively. During the spalling process, the top silicon layer and a 10-μm-thick silicon film from below the BOX are removed. The residual layer is then removed using chemical etching. After the spalling process, an Al layer is deposited on top of the BOX and a flexible tape is applied to serve as the new mechanical support for the released circuitry. Finally, the original handle and the Ni stressor layers are removed to provide access to the top fabricated circuits. Figure 2.3.3b [24] shows a cross-sectional TEM of the released film and the integrated circuits, where the circuits feature a 6-nm-thick silicon layer and a transistor with a 30-nm gate length and a 100-nm pitch. The obtained circuits were tested before and after the spalling process.

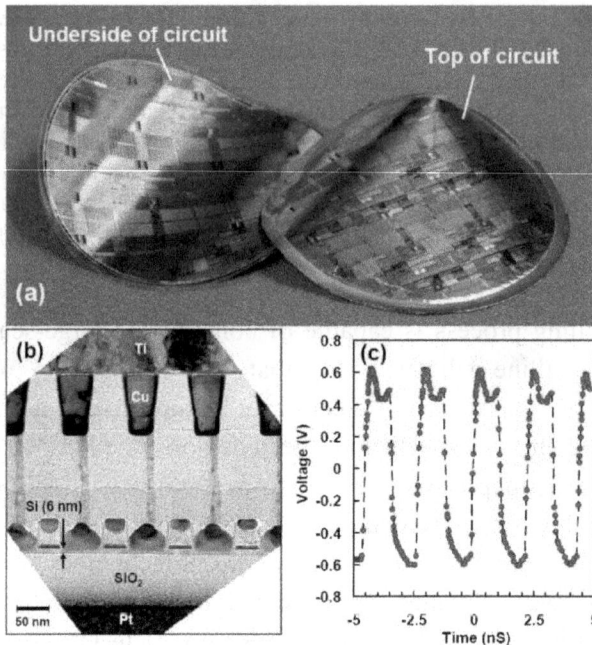

Figure 2.3.3: Reprinted with permission from [24]. (a) Flexible SOI-based CMOS circuits formed by controlled spalling. (b) Cross-section transmission electron microscope (X-TEM) image of the flexible circuit with the stressor and handle layers still intact. The Pt was deposited on the underlying SiO2 (BOX) layer during sample preparation. The Ti layer at the top of the image is the thin adhesion layer used before deposition of 6μm Ni stressor layer. The image shows the aggressive dimensional scaling of the transistors in these circuits as well as the absence of any defects induced during removal. (c) Voltage versus time waveform from a 100 stage ring oscillator illustrates the retained functionality of complex integrated device arrays after CST. The measured delay per stage of 11.4 ps from the flexible circuit is similar to measurements taken from the rigid circuit before spalling.

To examine the complete characteristics of the released devices [25], hundreds of fabricated transistors, ring oscillators, and static random access memories (SRAMs) were measured before and after the release process. Figure 2.3.4 [25] shows the comparison for transistors before and after spalling. Although results show a negligible change to transistor characteristics, a small variation appears in device performance following a shift in the threshold voltage ($V_{th}$) of 30 mV for both n- and p-FETs. Potentially this change is caused by the introduction of positively charged fixed charges during the bonding process. Note that the Ni stressor layer used during the spalling process causes the released circuits to be subject to residual compressive stress. However, this parasitic residual stress did not seem to impact device performance.

Comparing the position of the alignment marks between bulk and released samples identified residual strain between 0.15 and 0.18%. To characterize one of the basic components for system-on-chip, SRAM, comparison between released and bulk devices are shown in Figure 2.3.5.

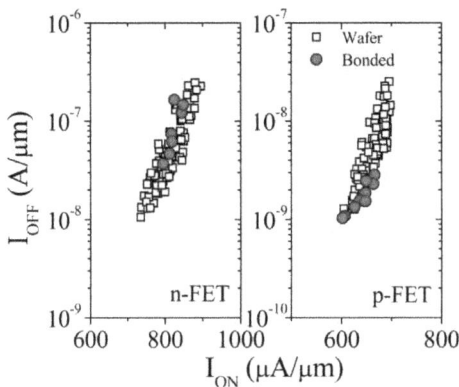

Figure 2.3.4: Reprinted with permission from (25). Copyright (2012) American Chemical Society. Comparison of the n- and p-FET device performance before and after the layer transfer. The equivalency of the device performance was confirmed by bonding the flexible sample in Figure 1d rigidly onto a silicon handle wafer prior to electrical measurements, eliminating the detrimental influence of the probe pressure on the electrical characteristic of the flexible devices.

The excellent subthreshold characteristics and the elimination of random dopant fluctuation provided by ultra-thin-body transistors allow for the reliable operation of SRAM cells with small power supplies. Figure 2.3.5a [25] shows the butterfly curves for the characterized flexible SRAM cells.

The ring oscillator is another circuit of interest for application to high-performance electronics. The obtained results show that the stage delay for 100-stage ROs is around 16 ps for flexible and bulk devices at a supply voltage ($V_{dd}$) of 0.9 V. Also, to test the stability of the fabricated flexible circuits under bending conditions, bending tests were performed using different cylinders with curvature radii ranging from 6.3 to 15.8 mm.

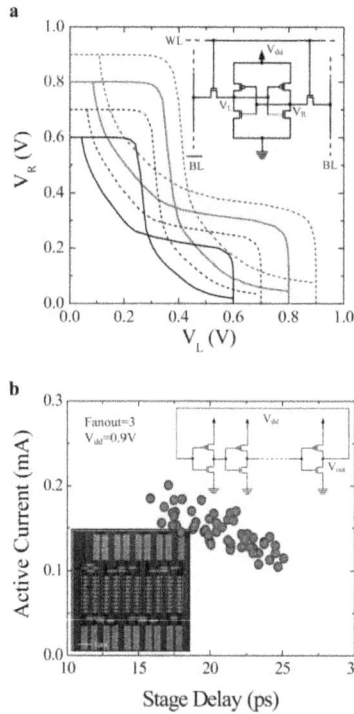

Figure 2.3.5: Reprinted with permission from (25). Copyright (2012) American Chemical Society. Ultradense high-performance flexible memory and ring oscillator circuits. (a) The representative butterfly curves for a flexible SRAM, showing good symmetry down to Vdd of 0.6 V. The inset shows the schematic illustration of a 6T-SRAM cell. (b) Stage delay characteristics of the flexible ring oscillators, indicating a record stage delay of = 16 ps. The top and bottom insets illustrate the circuit diagram of a multistage ring oscillator and the SEM image of a flexible 100-stage ring oscillator taken from the backside of the sample through the BOX, respectively.

Figure 2.3.6 [25] shows the transfer characteristics of an nFET under different tensile bending conditions. The obtained results show no performance degradation in the tested devices and only a small shift in the threshold voltage of 35 mV was observed at a banding radius of 6.3 mm. Finally, Figure 2.3.6c shows the consistency in the performance of the fabricated devices under different cycles of bending for a radius of 6.3 mm [24, 25].

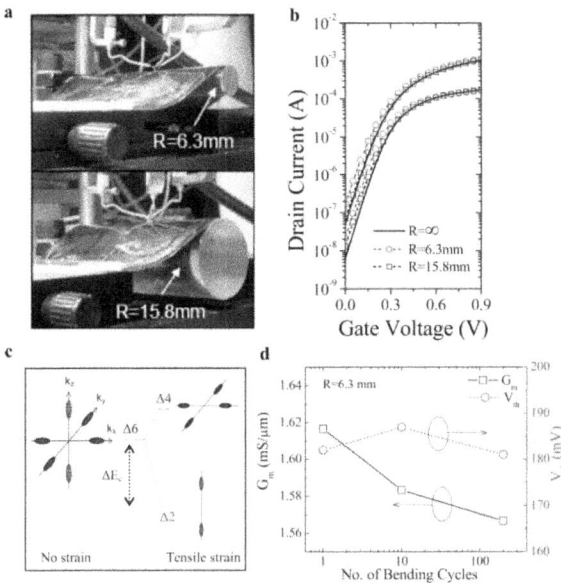

Figure 2.3.6: Reprinted with permission from (25). Copyright (2012) American Chemical Society. Bending stability of the flexible circuits. (a) Photograph of a flexible circuit under tensile bending tests at two different radii of curvature. (b) Transfer characteristics of an n-FET with a channel length of 20 nm under different bending conditions. (c) The bending tensile strain causes the 6-fold symmetry of the silicon conduction band to break, resulting in a shift in the threshold voltage of the n-FETs. (d) No noticeable change in device properties was observed during the repeated bending test at R = 6.3 mm.

This section concluded that spalling is a suitable technology for fabricating thin silicon circuitry. The advantage of simplicity and cost-effectiveness presented by this methodology presents a reliable solution for their integration into high-performance electronics on flexible platforms. Although complete 300-mm wafers can be peeled using the spalling process, in most the cases, only selected areas of the substrate need to be flexed due to design considerations and cost implications. The option of peeling complete or partial surfaces from the "mother" water using spalling makes it a process that can be used to produce high-performance electronic systems in a cost-effective and simplified manner.

## 3. High-performance 3D Nanoelectronics on Flexible Monocrystalline Silicon [35 – 38]

Since the introduction of CMOS technology for consumer electronics in 1974, device scaling has allowed the integration of billions of transistors into centimeter square chips. At present, CMOS technology is reaching a point where further scaling is not possible due to device degradation in terms of leakage current and short-channel effects. For this reason, FinFET architecture, the most advanced architecture for future CMOS-based circuitry has become the best solution to reduce scaling constraints [39]. FinFET, a member of the multi-gate field-effect transistor, consists of non-planar 3D channel devices. In FinFETs, the channels of the devices are vertically aligned in arrays of ultra-thin silicon pillars (Fins) that are surrounded by multiple gates. Intel, a giant in the semiconductor industry, has adopted FinFET technology for the fabrication of 22-nm-node microprocessors [40]. The advantages of FinFETs when compared to standard planar CMOS technology include the elimination of short-channel effects, enhanced drive current, and compatibility with fully depleted SOI wafers. In addition, SOI-based devices exhibit a lower subthreshold leakage current due to carrier confinement in the thin silicon body of the transistors [41, 42]. As the global demands for high-performance electronics increases, future architectures must evolve to exploit advantages and reduce side effects such as increased stand-by power consumption. For example, flexible devices have the potential to advance currently rigid wearable electronics towards becoming flexible to the point that they are able to adapt to the topography of the human body. However, to date, devices lack the required performance levels needed to satisfy the demands of consumer electronics. For this reason, and as described in the previous section, several groups have made efforts to convert rigid high-performance electronics into flexible high-performance electronics. The three main approaches described in Section 2 show numerous advantages in terms of performance and ease of fabrication. However, the transfer method championed by Prof. John Rogers et al. to fabricate silicon-based electronics from NRs and NMs released from a source substrate also presents some disadvantages in terms of long-established fabrication processes [1-13]. For example, high-resolution lithography and high-temperature-based depositions are commonly used in the fabrication of today's high-performance devices; these, however, are not compatible with plastic substrates due to their limited flatness and temperature instability. Meanwhile, Burghartz et al. introduced the unexplored processes of anodic etching [14,15], which they found limited bendability in the final devices and were associated with high-processing costs related to epitaxial silicon growth that prevents a cost-effective fabrication of high-performance electronics with this method. Finally, before the spalling process can be used to produce high-density flexible integrated systems we need to resolve its limited flexibility, use of costly substrates, and residual strains [24-26]. For this reason, in this section, we show how simple, industry compatible processes can be used to transform rigid 3D FinFET electronics into flexible ones, while maintaining the high-performance benefits of this advanced architecture.

## 3.1. High-Performance 3D FinFET Fabrication [35]

Fabrication starts with standard 90-nm-thick SOI wafers and 125-nm-thick buried oxide with a p-type silicon body ($2 \times 10^{15}$ atoms/cm$^3$) with no additional channel doping. High-resolution lithography and dry/wet etch techniques are used to pattern the narrow fins. Figure 3.1.1 [35] depicts the process for gate stack integration.

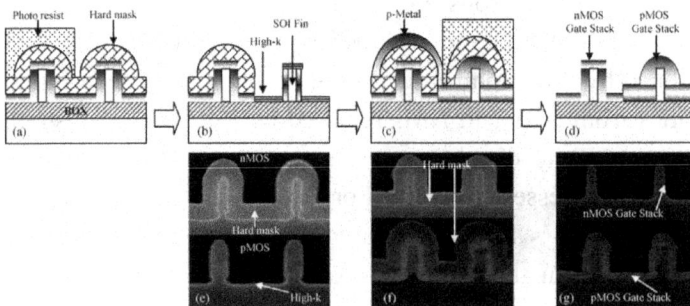

Figure 3.1.1: © [2010] IEEE. Reprinted, with permission, from [35]. Process flow for TWF CMOS FinFETs gate stack integration—(top) schematics and (bottom) SEM. (a) NMOS gate stack with a sacrificial hard-mask deposition and pattern lithography. (b) and (e) Hard-mask dry etch, photoresist removal followed by metal wet etch. (c) and (f) p-metal, sacrificial hard-mask deposition, and lithography pattern to dry etch of hard mask followed by photoresist removal and p-metal wet etch. (d) and (g) Hard-mask etch to form NMOS & PMOS gate stacks.

To fabricate tuned-work-function high-k/metal gate transistors, a wet etch process was incorporated as part of the gate stack definition. After a 3-nm ALD high-k dielectric is deposited on the wafers, a thin (3 nm) titanium nitride (TiN) layer is deposited followed by a poly-Si hard mask. High-resolution photolithography is used to pattern openings over the PMOS areas, and, high-density plasma etching with a 1:1 lateral to vertical etch selectivity removes the poly-Si hard mask from the PMOS region. An ammonia-based ($NH_3$) ash and NMP cleaning remove the photoresist, and wet etching in RCA (DI-water:$H_2O_2$:$NH_4OH$) removes the thin NMOS ALD TiN prior to deposition of the thick PMOS ALD TiN layer. A second poly-Si hard mask is deposited and patterned with photoresist to expose the buried NMOW areas. Using the same process as that used for PMOS devices, the hard mask is removed from NMOS areas. RCA cleaning selectively removes the remaining thick (PMOS) TiN layer from covering the NMOS devices. Oxide- and OH-based wet etch removes the remaining poly-Si hard mask from both PMOS and NMOS devices, and standard poly-Si deposition and anisotropic dry etching patterned the gate. Next, a nitride spacer is deposited and then patterned using highly anisotropic dry etching techniques. Finally, source and drain implantation creates highly doped regions of the FinFETs and NiSi and AlSi metallization is used to complete the fabrication flow of the

transitors. The fabricated devices feature 20-nm fins (width =
27 nm and height = 85 nm) in arrays of 2 and 20 fins. Figure
3.1.2 [35] shows TEM imaging of the fins and fabricated
devices.

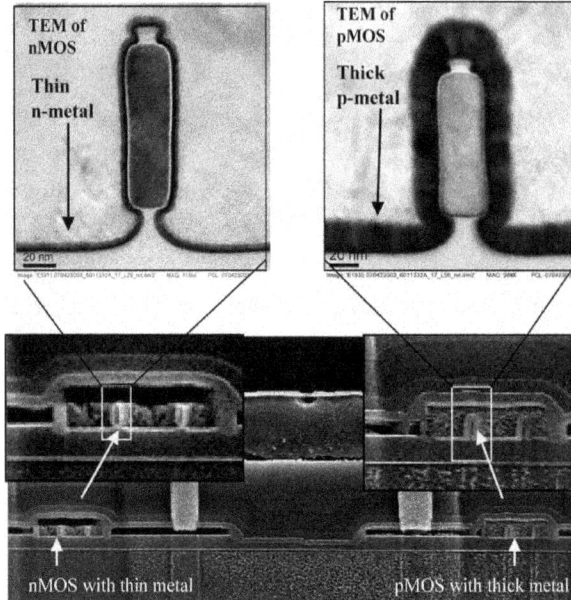

Figure 3.1.2: © [2010] IEEE. Reprinted, with permission, from [35].
SEM image of fully integrated TWF high-k/metal gate CMOS FinFETs
& magnified TEM images of fins with high-k/metal gate stack.

## 3.2. Deterministic Pattern-Based FinFET Release [36]

Figure 3.2.1 [36] shows how the flow followed to release the
fabricated FinFET devices. Once the FinFETs are fabricated
using the process described in the previous subsection, they are
diced into 2.5 cm × 3 cm pieces for individual processing. Each

piece is spin coated with thick photoresist (PR) (nLOF 2070) to obtain a PR of 7 μm. Next, the dies are processed using research photolithography to create etch hole patterns in the inactive transistor areas. The distance between holes is determined based on two basic elements. First, selectivity between silicon and BOX need to be high to use the BOX layer as an etch stop. And second, the BOX layer needs to be thick enough that it will maintain its integrity during the isotropic etching of silicon. Since silicon to oxide etch in xenon-diflouride has a selectivity higher than 1000:1, a 125-nm-thick BOX layer required to support the release process when holes are separated by 150 μm. Next, the ILD comprising 700-nm $Si_3N_4$ and 300-nm $SiO_2$ is removed from the area of the holes with high-density plasma to expose the BOX at the bottom of the trenches. Once the BOX is removed with a highly directional low-etch-rate ICP RIE process until the thick silicon body is vented the wafers are placed in an $XeF_2$ silicon isotropic etchant to remove the silicon from the bottom of the BOX and create caves. Once these caves meet, the top Si thin body and the BOX are completely released from the bulk substrate. The released layer is then transferred to a thin (125 μm) Kapton handle substrate, which has a thin partially cured poly-dimethylsiloxane layer to enhance adhesion between the Kapton sheet and the peeled devices. Finally, the PDMS is cured for 24 hours at room temperature and atmospheric pressure to avoid damaging the transferred devices due to unwanted thermal deformation. The released pieces are 1000 FinFET n- and p-type devices.

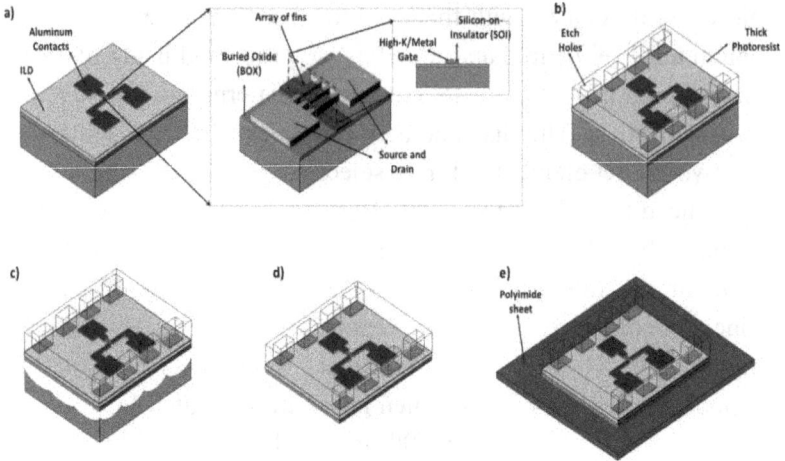

Figure 3.2.1: [36]. Copyright (c) [2014] [2014 WILEY-VCH Verlag GmbH & Co. KGaA, Weinheim]. Schematic flow for silicon fabric release containing FinFETs. a) Initial FinFET device fabricated on 90 nm thick SOI, aluminum contact pads, and 850 nm ILD. Projection: FinFET channel, source and drain and gate stack, b) Spin coat of thick (7μm) photoresist and hole patterning, c) Cavern formation beneath BOX due to XeF$_2$ etchant, d) Released fabric (1 μm) after 600 cycles in XeF$_2$ isotropic etchant, e) Transferred devices to 125 μm thick polyimide sheet spin coated with thin PDMS adhesion layer (≈1 μm).

## 3.3. Results Obtained With Deterministic Pattern-Based Release [36]

The behavior of the fabricated bulk and released FinFETs were obtained by first extracting the $I_d$-$V_g$ characteristics of the transistors. Direct properties such as saturation current, $I_{on}/I_{off}$ ratio, and gate-leakage current density are extracted from the obtained curves. The values procured for FinFET devices of 250-nm gate length and 3.6-μm channel width were: a saturation current of 549 μA/μm with a standard deviation of 13.96 μA/μm for NMOS devices driven at $V_{GS}$ = 1.5 V and $V_{DS}$ = 1 V. In

PMOS devices, a saturation current of 110 $\mu A/\mu m$ with a standard deviation of 1.52 $\mu A/\mu m$ is measured when driven at – 1.5 V for $V_{GS}$ and –1V for $V_{DS}$. The subthreshold current was found to be 58.48 $\mu A/\mu m$ and 7.73 $\mu A/\mu m$ at $V_{DS} = 50$ mV and – 50 mV for N- and PMOS, respectively. In the case of NMOS devices, the $I_{on}/I_{off}$ ratio was 4.6 decades and for PMOS 4.78 decades. For NMOS devices, gate-leakage current density was found as 3.6 $A/cm^2$ at $V_{GS} = 1.5$ V. PMOS devices yielded a gate-leakage current density of 1.18 $A/cm^2$ when driven at a $V_{GS} = -1.5$ V. Extraction of the main electrical parameters started by obtaining the threshold voltage from the $I_D$-$V_G$ curve using the linear extrapolation method [43]. The $V_{th}$ can be determined by

$$V_{th} = V_{GS_0} - \frac{V_{DS}}{2} \quad (3.3.1)$$

Here, $V_{GS0}$ is the intercepting point of the linear extrapolation with the $V_G$ axis (x-axis) of the $I_D$-$V_G$ plot. The obtained value for NMOS devices was 0.345 V and 0.713 V for PMOS FinFETs. Next, effective mobility was extracted at low drain voltages (50 mV) using the following equation:

$$\mu_{eff} = \frac{L}{W} \frac{g_d}{C_{ox}(V_{GS} - V_{th})} \quad (3.3.2)$$

Here, L is the gate length of the device, W is the channel width, $V_{GS}$ is the gate to source voltage, $C_{ox}$ is the gate capacitance, $V_{th}$ is the threshold voltage, and $g_d$ is the device transconductance, which can be determined by

$$g_d = \frac{\partial I_{Deff}}{\partial I_{DS}} | V_{GS} = constant \quad (3.3.3)$$

In this case, $I_{Deff}$ is the effective drain current when half the gate leakage is added to the drain current value. This addition is required when extremely scaled gate oxides are used for device fabrication; in this case gate oxide was (3 nm $HfO_2$). $I_{Deff}$ is determined as

$$I_{Deff} = I_D + \frac{I_G}{2} \quad (3.3.4)$$

Using previous equations, the mobility of NMOS devices was calculated at 141.53 $cm^2V^{-1}s^{-1}$ and at 3.74 $cm^2V^{-1}s^{-1}$ for PMOS FinFETs. Although this mobility value for PMOS devices may seem low compared to previously reported values [44], note that the release process did not affect the effective mobility values when comparing released with bulk samples. Finally, the subthreshold swing was found to be 80 $mVdec^{-1}$ for NMOS devices and 70 $mVdec^{-1}$ for PMOS transistors. We measured drive current reduction by comparing values for flexible and bulk devices as only 7 and 12% for NMOS and PMOS transistors, respectively.

As previously reported [36], this reduction can be explained by the amount of residual stress contributed by the thick ILD, which becomes more significant when devices are in the release state due to the lack of a bulk support substrate. In the case of "*off*" state current, there was no significant change when comparing released with bulk devices. For NMOS, $I_{off}$ was found to be 13.79 nA/μm for released devices and 15.63 nA/μm for bulk transistors. In the case of PMOS, $I_{off}$ was calculated as 1.82 nA/μm for flexible transistors and 3.54 nA/μm for bulk devices. The difference in $V_{th}$ for released and bulk devices was obtained as 6% for NMOS and 8% for PMOS. Finally, the change between released and bulk devices was found to be as low as

3.4% for NMOS transistors and 2.8% for PMOS transistors. Figure 3.3.1 [36] shows the comparison between released and bulk FinFETs.

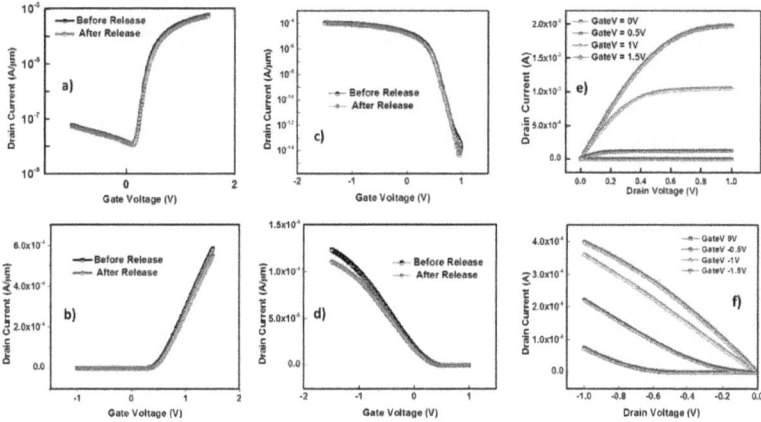

Figure 3.3.1: [36]. Copyright (c) [2014] [2014 WILEY-VCH Verlag GmbH & Co. KGaA, Weinheim]. Electrical characteristics comparison between P- and NMOS devices on before and after release silicon fabric (L = 250 nm, W = 3.6 μm, $V_{ds}$ = 1 V). a) $I_D$ –$V_G$ transfer characteristics in logarithmic scale for NMOS. b) $I_D$ –$V_G$ transfer characteristics in linear scale for NMOS. c) $I_D$ –$V_G$ transfer characteristics in logarithmic scale for PMOS. d) $I_D$ –$V_D$ transfer characteristics in linear scale for PMOS. e) $I_D$ –$V_D$ curves for NMOS. f) $I_D$ –$V_D$ curves for PMOS.

The mechanical characterization of devices consists of obtaining the performance of the released transistors under different bending conditions and applied stress. The minimum bending radius at which the devices fail due to interconnection damage was found to be 5 mm. Devices were characterized electrically by using different bending radii until the 5-mm limit was achieved. At the limit radius, the top surface of the released sample is subject to 0.0125% strain in the longitudinal direction

166

FLEXIBLE SILICON NANOELECTRONICS

of the channel. Figure 3.3.2 [36] shows the electrical characteristics of the released devices under different nominal strain values for banding radii of 5, 10, 15, 20, 30, and 50 mm. Since subthreshold swing is one of the key parameters for FinFET devices, its change with respect to the applied strain was studied, and a maximum change of only 3% was obtained when comparing flat with bent devices. In addition, the $I_{off}$ current increased by 10% for NMOS devices and by 9% increment for PMOS devices; $I_{on}$ decreased by 11% for NMOS devices and by 8% reduction for PMOS devices. Finally, positives shifts were observed for $V_{th}$: 3% for the $V_G$ of NMOS devices and 2% in the x-direction for PMOS devices.

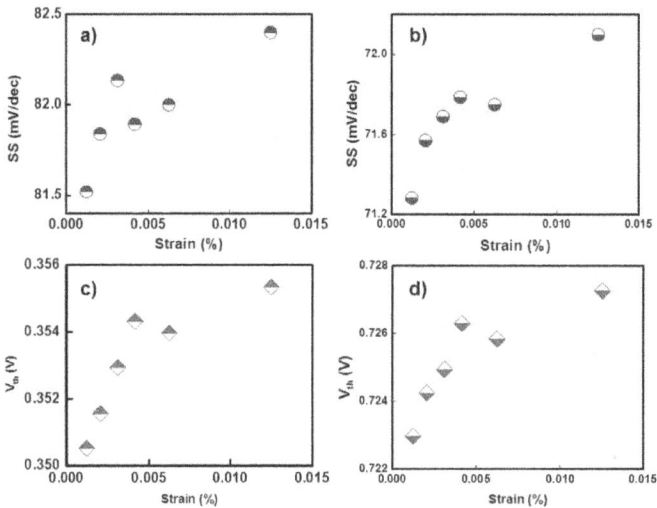

Figure 3.3.2. [36]. Copyright (c) [2014] [2014 WILEY-VCH Verlag GmbH & Co. KGaA, Weinheim]. Device performance dependence on applied nominal strain. a-b) Threshold voltage ( $V_{th}$ ) behavior under different strain conditions. c-d) Sub-threshold slope (SS) behavior under different strain conditions.

## 3.4. Soft Back-Etch FinFET Thinning Process [37]

As in the deterministic pattern release process, fabricated FinFETs were diced into 2.5 cm × 3 cm pieces to allow for the processing of single chips. Once prepared, the individual chips were spin coated with a thick layer of photoresist (7 μm) to protect the front surface from damage caused by the soft back-etch process. A 7-μm thick photoresist was chosen because thicker polymers provide better thermal isolation between the cooled stage of the RIE tool and the etched chip, and hence protect the top devices from performance degradation due to unwanted temperature effects. The chips are turned upside down and placed on a carrier 4'' wafer for easier handling, and the back surface of the chips is etched using a highly anisotropic etching process (Bosch) to reduce Si thickness and achieve flexibility. Bosch etching is a type of dry, plasma-based etching where high-density reactive plasma is created in the etching chamber and is accelerated towards the etching surface. During Bosch process, two different reactions occur, the first one is due to highly reactive $SF_6$ species, which react with Si atoms to produce $SiF_4$ molecules in a gaseous state, which are then pumped out of the system with the aid of a vacuum pump. The second reaction is due to the non-reactive nature of $C_4F_8$ which coats the silicon wafer with a fluorocarbon-based polymer to protect the sidewalls of the features during subsequent etch steps. To prevent over etching of the back surface and damage to the top surface, the back etching process is divided into four steps. In the first step, high-density plasma and a fast etch rate thin the chip from 800 μm to 200 μm. Then, using the same etching process but a different etch time, the back surface of the chips is etched 50 μm at a time until the target thickness < 50 μm is achieved. The Bosch process is performed at ☐20 ˚C, and comprises an etch step and a deposition step. The etch step lasts

7 s, with 1300 watts ICP power, 30 watts RF power, a pressure of 30 mTorr, and a gas flow of 5 sccm $C_4F_8$ and 120 sccm $SF_6$. This etch steps etches silicon at a high etch rate due to highly reactive fluorine-based ions and high-density plasma. The deposition step lasts 5 s, with 1300 watts ICP power, 5 watts RF power, a pressure of 30 mTorr, and a gas flow of 100 sccm $C_4F_8$ and 5 sccm $SF_6$. During deposition, low forward power is used to enhance thin polymer deposition on the wafers and prevent ions from etching the surface. To control the thickness of the thinned chip, a measurement step is performed between each etching step. These measurements use two different profiling tools. The first tool is a force-based profiling process that uses a micro-tip to measure the step size between two different surfaces; the second tool is a light-based profiling system that uses reflected light to measure the step size between two adjacent surfaces. Finally, once the target thickness is achieved (< 50 μm), the thick layer of photoresist is removed from the top surface and the finalized flexible chip can be transferred to a polymer-based carrier substrate or be supported by the thin bulk while measurements are conducted. Figure 3.4.1 [37] shows how silicon FinFETs are thinned down using the developed soft etch-back process.

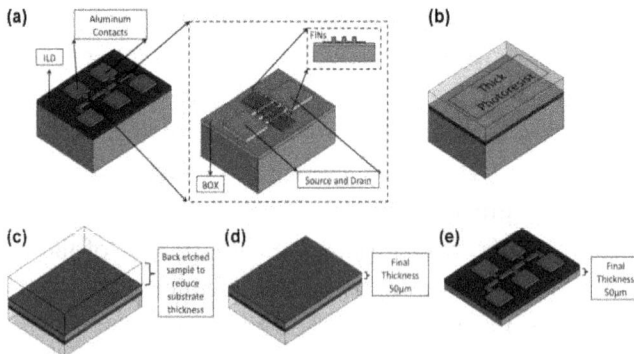

Figure 3.4.1: Reprinted with permission from (37). Copyright (2014) American Chemical Society. Fabrication process flow: (a) fabricated FinFET devices on standard 90 nm SOI with 150 nm BOX; (b) PR coating for chip-protection during back etch process; (c) FinFET die etched back using back grinding technique; (d) FinFET devices on flexible silicon substrate (50 μm thick); and (e) PR removal and final device testing.

## 3.5. Results Obtained Using the Soft Etch-Back Thinning Process [37]

Following the same procedure as described in subsection 3.3, the electrical characteristics of bulk and flexible FinFETs were obtained by first extracting the direct results from the $I_D$-$V_G$ curves. Saturation currents were calculated for NMOS devices with a gate length of 1 μm at 83 μA/μm and for PMOS transistors (gate length = 250 nm) at 383 μA/μm. Figure 3.5.1 [37] shows a comparison between released and bulk devices. Because PMOS devices had shorter gate lengths, their characteristics were studied extensively under different bending conditions for tensile and compressive stress.

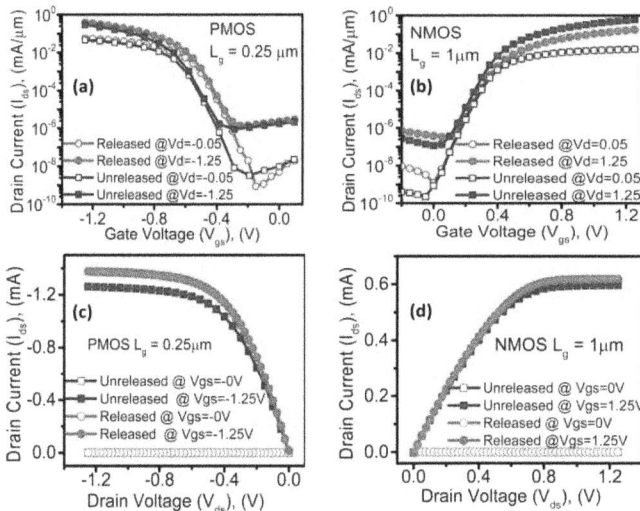

Figure 3.5.1: Reprinted with permission from (37). Copyright (2014) American Chemical Society. Unreleased & released transfer characteristics: (a) PMOS FinFET transfer curves ($L_g$ = 250 nm, W = 3.6 µm); (b) NMOS FinFET ($L_g$= 1 µm, W = 3.6 µm). Unreleased and released output characteristics: (c) PMOS FinFET transfer curves ($L_g$ = 250 nm, W = 3.6 µm) and (d) NMOS FinFET ($L_g$ = 1 µm, W = 3.6 µm).

To complete the characterization and to understand the changes introduced to device performance from the thinning process and bending conditions, the electrical characterization followed the same procedure as in subsection 3.3 to obtain the key parameters in different states. The obtained $V_{th}$ using linear extrapolation method was −0.556 V in the linear region and −0.474 in saturation. Equations 3.3.2, 3.3.3, and 3.3.4 were used to extract mobility at 102 $cm^2V^{-1}s^{-1}$. The $I_D$-$V_G$ curve was used to calculate the subthreshold swing at 63 mVdec$^{-1}$. The drain-induced barrier lowering (DIBL) for the released sample was measured at 68 mV/V, and the $I_{on}$/$I_{off}$ ratio was 5 decades with no significant changes during bending cycles. The released sample showed peak transconductance at 0.33 mS, and finally, gate-leakage current density was found to be 0.9 $mA/cm^2$. Figure 3.5.2 [37] shows the transfer and output characteristics of the released samples at different bending radii.

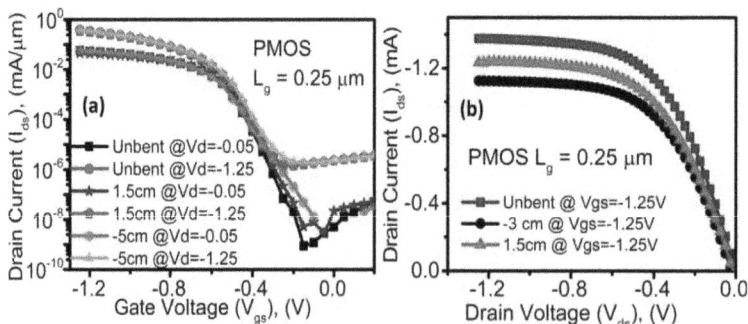

Figure 3.5.2: Reprinted with permission from (37). Copyright (2014) American Chemical Society. Unbent and bent (a) transfer and (b) output characteristics in flexible PMOS FinFETs ($L_g$ = 250 nm, W = 3.6 μm).

We conducted the mechanical characterizations under compressive and tensile stress bending conditions to find a minimum tensile radius of 1.5 cm and a minimum compressive radius of 3 cm. A maximum change in saturation current was found at 7.6% with a 1.5-cm tensile bending radius, and a maximum change in SS was found at 10.3% with a 5-cm tensile bending radius. $J_G$ demonstrates a maximum two order of magnitude ($10^2$) change at a compressive bending radius of 3 cm. The maximum change in DIBL was found to be 3.8 times at a 5-cm compressive bending radius, while $V_{th}$ had a maximum change of 7.4% and 5.1% in linear and saturation regions, respectively. The maximum change in transconductance $g_m$ was found to be 36.4% at a compressive bending radius of 3 cm, and the $I_{on}/I_{off}$ ratio showed a maximum change of 15.9% at a tensile bending radius of 5 cm. Finally, the maximum change in peak mobility was found to be 29.6% at a tensile bending radius of 1.5 cm. Figures 3.5.3 [37] and 3.5.4 [37] show the change in the described transistor characteristics under different compressive and tensile bending states.

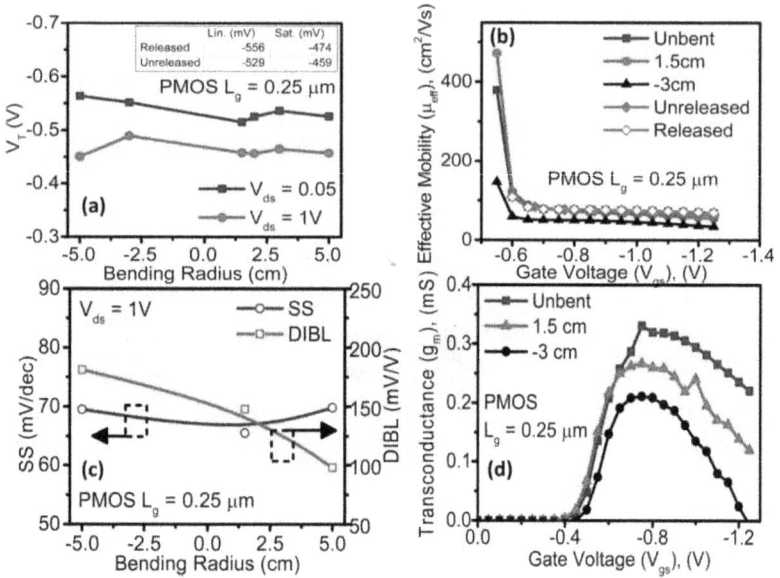

Figure 3.5.3: Reprinted with permission from (37). Copyright (2014) American Chemical Society. (a) Threshold-voltage ($V_t$) vs bending radii in flexible PMOS FinFETs ($L_g$ = 250 nm, W = 3.6 μm); (b) mobility variation due to bending; (c) subthreshold swing and DIBL vs bending radii and (d) bending effect on transconductance.

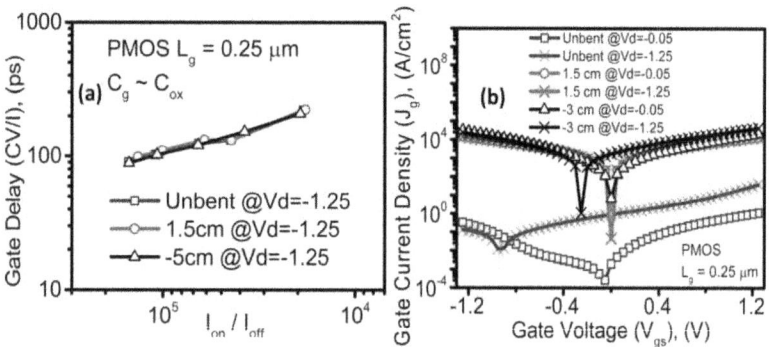

Figure 3.5.4. Reprinted with permission from (37). Copyright (2014) American Chemical Society. Effects of bending radii on (a) gate-delay and (b) leakage current density.

## 3.6. High-Temperature Measurements of Flexible FinFETs [38]

The high-temperature performance of released and bulk devices was studied to understand how the devices behave under harsh environments. The electrical characteristics of the transistors were analyzed on a controlled temperature stage between 25 °C and 150 °C. Figure 3.6.1 [38] shows the electrical performance of the released FinFETs (gate length = 250 nm) under different temperature conditions, with the transistors biased with a $V_{DD}$ = –1 V. From Figure 3.6.1a, which shows the $I_D$-$V_G$ curve at different temperatures, we note that the threshold voltage decreases due to a shift in the Fermi level as the temperature of the device increases. Since a reduction in $V_{th}$ will increase the drain current and a reduction in carrier mobility will decrease drain current [35, 45], these opposing effects lead to unique characteristics of devices that have a zero temperature coefficient (ZTC). To understand the behavior of the devices in terms of subthreshold swing, we plotted the $I_D$-$V_G$ plot in a semi-logarithmic scale; Figure 3.6.1b shows how the SS increases in direct proportionality to temperature, and how in the range of 150 °C, FinFETs undergo low leakage current. Figure 3.6.1c shows that transconductance is reduced with increasing temperature, which is easily explained by understanding that at elevated temperatures, phonon scattering increases, causing a reduction in carrier mobility [46, 47].

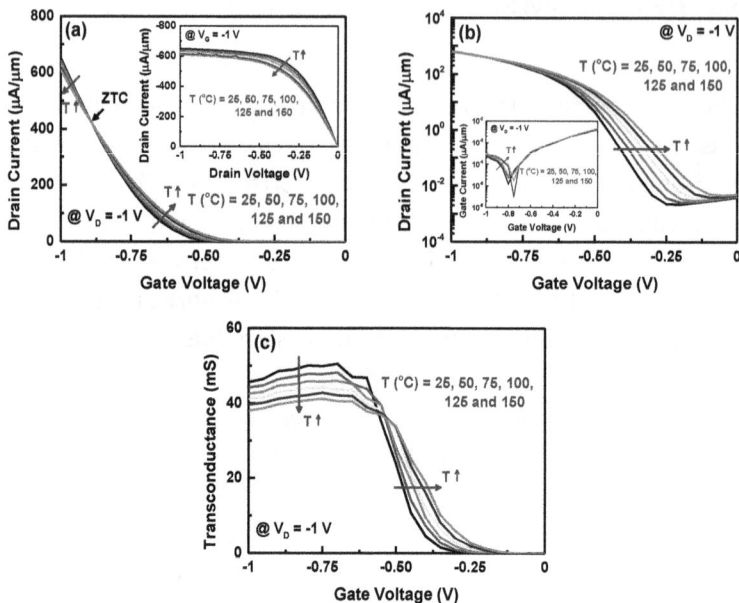

Figure 3.6.1. © [2014] IEEE. Reprinted, with permission, from [38]. Flexible 250-nm LG FinFET characteristics from room (25 °C) to high (150 °C) temperatures. (a) Transfer plots ($I_D$–$V_G$) in linear scale. Inset: output plots ($I_D$–$V_D$) for VG = −1 V. (b) Transfer plots ($I_D$ –$V_G$) in semilogarithmic scale. Inset: $I_G$ versus $V_G$ curves. (c) Transconductance (gm) curves as a function of $V_G$ for same device and temperatures values ($V_D$ = −1 V).

A comparison between mechanical states at 25 °C and 150 °C was made to explain how higher temperatures affect released and unreleased devices. Figure 3.6.2 [38] shows that the devices behave similarly and almost no variation in $V_{th}$ or on-state current is observed. The overlap between the curves shows that the subthreshold swing is not altered by the release process. Also, a nanoampere-sized increase in $J_G$ takes place with increasing temperature. Finally, transconductance can be seen to superpose without a marked change between released and unreleased samples. Therefore, it appears that exposure to high temperatures (150 °C) has a negligible effect on the high-performance nature of devices even after the release process.

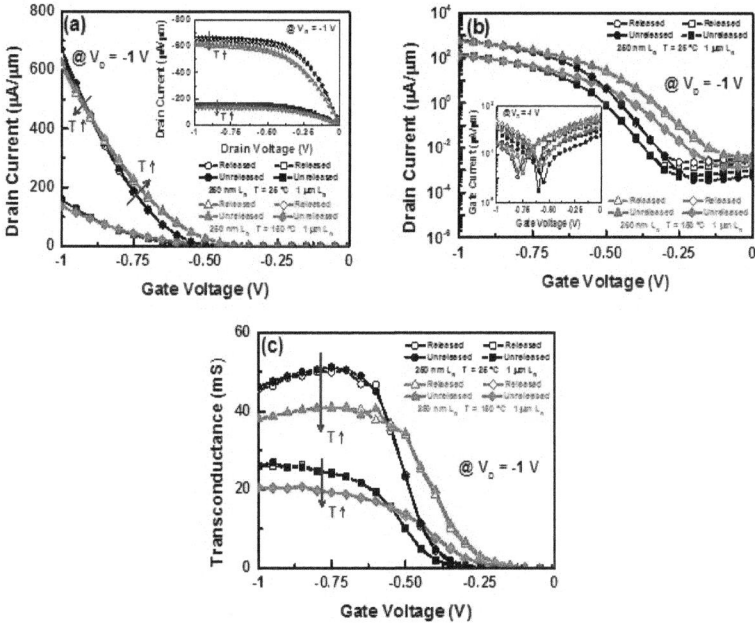

Figure 3.6.2. © [2014] IEEE. Reprinted, with permission, from [38]. Unreleased and released FinFET. (a) $I_D$ –$V_G$ plots in linear scale for 25 °C and 150 °C. Inset: $I_D$–$V_D$ plots for same devices and temperatures. Unreleased and released FinFET. (b) $I_D$–$V_G$ plots in semilogarithmic scale for 25 °C and 150 °C. Inset: gate current $I_G$ versus $V_G$ curves for same devices and temperatures. (c) Transconductance versus $V_G$ for unreleased and released FinFETs at 25 °C and 150 °C.

A quantitative analysis of the key characteristics of released and unreleased FinFET devices exposed to different temperatures is presented in Figures 3.6.3, 3.6.4, 3.6.5, and 3.6.6 [38].

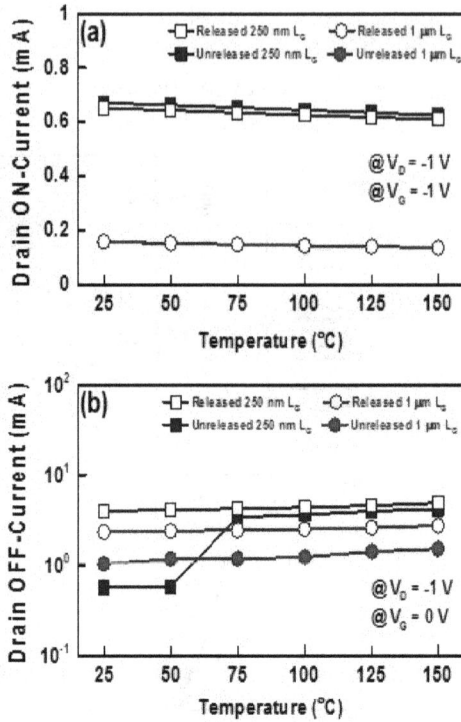

Figure 3.6.3: © [2014] IEEE. Reprinted, with permission, from [38]. (a) Drain ON-current (measured at $V_D = -1$ V and $V_G = -1$ V) and (b) drain OFF-current (measured at $V_D = -1$ V and $V_G = 0$ V) of unreleased and released FinFETs as a function of temperature.

Figure 3.6.3a shows on-state current as a function of temperature for released and unreleased FinFETs. Since the gate bias is larger than that of the ZTC, the $I_{on}$ current decreases with temperature [48]; this degradation is due to reduced carrier mobility in the channel of the devices. Note that if the drain voltage is maintained below the ZTC, the on-state current will increase with temperature because decreasing $V_{th}$ dominates current behavior. Figure 3.6.3b shows a small increase in the off-state current of the transistors, which can be explained by a higher intrinsic carrier concentration that causes both diffusion and generation currents [49]. Figure 3.6.4a shows that the gate-leakage current behaves nearly linearly, but very small increases at high temperatures occur as a result of the high-k nature of the dielectric. A small increase in gate-induced drain leakage (GIDL) can be explained by band-to-band tunneling and band-to-defect tunneling [50]. Variations in GIDL of less than one order of magnitude between room- and high-temperature (150 °C) measurements can be seen in Figure 3.6.4b.

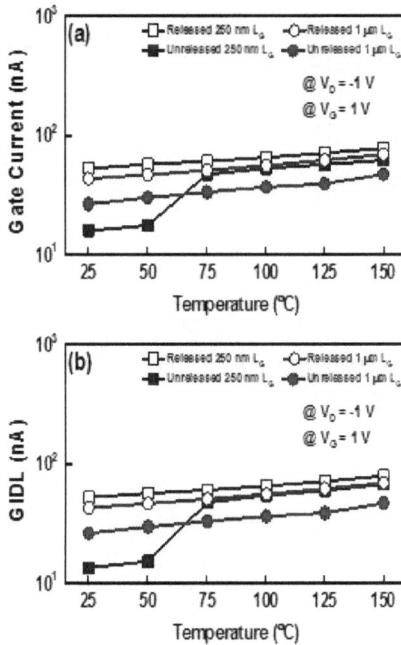

Figure 3.6.4. © [2014] IEEE. Reprinted, with permission, from [38]. Unreleased and released FinFET. (a) Gate-leakage current curves versus temperature. (b) GIDL versus temperature. Measured values at $V_D = -1$ V and $V_G = -1$ V. The curves are plotted in semilogarithmic scale.

$V_{th}$, SS, and mobility were extracted as a function of temperature for released and unreleased devices at low drain voltages (50 mV). To circumvent the effects produced by the increment in series resistance between the probes and the contact pads as temperature increases, $V_{th}$ and $\mu_0$ were extracted using the Y-function method [51]. In this method, the intercept between the Y-function and gate voltage yields the threshold voltage of the device, and the slope of the curve gives the $\mu_0$ parameter of the transistor. The variation in threshold voltage and the variation in SS as a function of temperature can be seen in Figure 3.6.5.

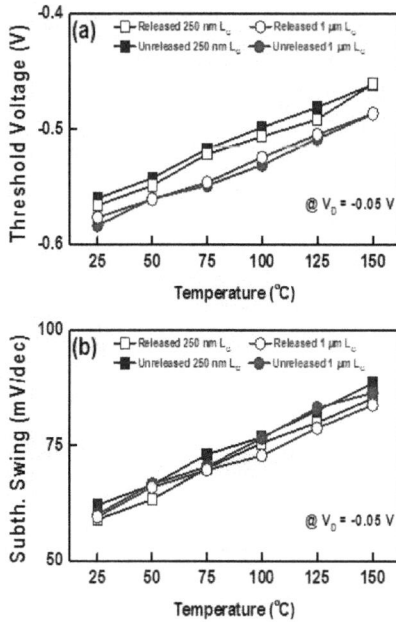

Figure 3.6.5. © [2014] IEEE. Reprinted, with permission, from [38]. Unreleased and released FinFETs. (a) $V_{th}$ variation with temperature. (b) SS variation with temperature. The values are extracted using Y-function in linear regime ($V_D = -0.05$ V).

Finally, the extracted mobility and its change with temperature can be seen in Figure 3.6.6. A small decrease in mobility is due to increased phonon scattering in response to increased temperature ($\mu \propto T^{-n}$) [52, 53]. In high-quality MOSFETs, surface roughness scattering occurs only at high vertical fields, whereas Coulomb scattering is negligible in undoped transistors operated at room temperature and above. Alternatively, in fabricated FinFETs (n = 1.7) the release process does not increase the defect density in the channel of the devices, hence preserving the high quality of the dielectric/channel interface.

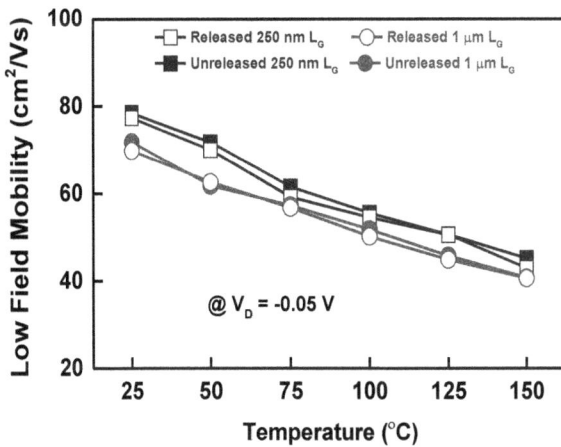

Figure 3.6.6: © [2014] IEEE. Reprinted, with permission, from [38]. Low-field mobility (extracted at $V_D$ = −0.05 V from the Y-function method) versus temperature for unreleased and released FinFET.

# References

[1] S. Mack, M. A. Meitl, A. J. Baca, Z.-T. Zhu, J. A. Rogers, Appl. Phys. Lett. 2006, 88, 213 101.

[2] E. Menard, K. J. Lee, D.-Y. Khang, R. G. Nuzzo, J. A. Rogers, Appl. Phys. Lett. 2004, 84, 5398.

[3] E. Menard, R. G. Nuzzo, J. A. Rogers, Appl. Phys. Lett. 2005, 86, 093 507

[4] Z.-T. Zhu, E. Menard, K. Hurley, R. G. Nuzzo, J. A. Rogers, Appl. Phys. Lett. 2005, 86, 133 507.

[5] J.-H. Ahn, H.-S. Kim, K. J. Lee, Z. Zhu, E. Menard, R. G. Nuzzo, J. A. Rogers, IEEE Electron Device Lett. 2006, 27, 460.

[6] Y. Sun, S. Kim, I. Adesida, J. A. Rogers, Appl. Phys. Lett. 2005, 87, 083 501.

[7] Y. Sun, E. Menard, J. A. Rogers, H.-S. Kim, S. Kim, G. Chen, I. Adesida, R. Dettmer, R. Cortez, A. Tewksbury, Appl. Phys. Lett. 2006, 88, 183 509.

[8] H.-C. Yuan, Z. Ma, M. M. Roberts, D. E. Savage, M. G. Lagally, J. Appl. Phys. 2006, 100, 013 708. b) H. Y. Li, L. H. Guo, W. Y. Loh, L. K. Bera, Q. X. Zhang, N. Hwang, E. B. Liao, K. W. Teoh, H. M. Chua, Z. X. Shen, C. K. Cheng, G. Q. Li, D.-L. Kwong, IEEE Electron Device Lett. 2006, 27, 538.

[9] Y. Sun, J. A. Rogers, Nano Lett. 2004, 4, 1953

[10] K. J. Lee, M. J. Motala, M. A. Meitl, W. R. Childs, E. Menard, A. K. Shim, J. A. Rogers, R. G. Nuzzo, Adv. Mater. 2005, 17, 2332.

[11] Kim, Dae-Hyeong, et al. "Flexible and stretchable electronics for biointegrated devices." *Annual review of biomedical engineering* 14 (2012): 113-128.

[12] Kim DH, Ahn JH, Choi WM, Kim HS, Kim TH, et al. 2008. Stretchable and foldable silicon integrated circuits. Science 320:507–11

[13] Ahn, Jong-Hyun, et al. "Heterogeneous three-dimensional electronics by use of printed semiconductor nanomaterials." *science* 314.5806 (2006): 1754-1757.

[14] Burghartz, Joachim N., et al. "Ultra-thin chip technology and applications, a new paradigm in silicon technology." *Solid-State Electronics* 54.9 (2010): 818-829.

[15] Burghartz, Joachim N., et al. "A new fabrication and assembly process for ultrathin chips." *Electron Devices, IEEE Transactions on* 56.2 (2009): 321-327.

[16] Peterson K. Silicon as a mechanical material. Proc IEEE 1982;70(5):420–57.

[17] Koyanagi M, Nakamura T, Yamada Y, Kikuchi H, Fukushima T, Tanaka T, et al. Three-dimensional integration technology based on wafer bonding with vertical buried interconnections. IEEE Trans El Dev 2006;53(11):2799–808.

[18] Jung E, Neumann A, Wojakowski D, Ostmann A, Landsberger C, Aschenbrenner R, et al. Ultra thin chips for miniaturized products. In: Proc. 2002 electronics components and materials conference, ECTC, 2002, p. 1110–3.

[19] Landesberger C, Klink G, Schwinn G, Aschenbrenner R. New dicing and thinning concept improves mechanical reliability of ultra-thin silicon. In: Proc int symp adv pack mat; 2001, p. 92–7.

[20] Swinnen B, Ruythooren W, DeMoor P, Bogaerts L, Carbonell L, De Munck K, et al. 3D integration by Cu–Cu thermo-compression bonding of extremely thinned bulk-Si die containing 10 lm pitch through-Si vias. In: Techn dig int el dev mtg (IEDM); 2006. p. 371–80.

[21] Dekker R. Sustrate transfer technology. PhD thesis, TU Delft, The Netherlands; 2 June 2004

[22] S. Armbruster, F. Schäfer, G. Lammel, H. Artmann, C. Schelling, H. Benzel, S. Finkbeiner, F. Lärmer, P. Ruther, and O. Paul, "A novel micromachining process for the fabrication of monocrystalline Si-membranes using porous silicon," in Proc. 12th Int. Conf. Solid-State Sens., Actuators Microsyst., Boston, MA, Jun. 8–12, 2003, pp. 246–249.

[23] T. J. Rinke, R. B. Bergmann, and J. H. Werner, "Quasi-monocrystalline silicon for thin-film devices," Appl. Phys., A Mater. Sci. Process., vol. 68, no. 6, pp. 705–707, Jun. 1999

[24] Bedell, Stephen W., et al. "Layer transfer by controlled spalling." *Journal of Physics D: Applied Physics* 46.15 (2013): 152002.

[25] Shahrjerdi, Davood, and Stephen W. Bedell. "Extremely flexible nanoscale ultrathin body silicon integrated circuits on plastic." *Nano letters* 13.1 (2012): 315-320.

[26] Rao, R. A., et al. "A novel low cost 25μm thin exfoliated monocrystalline Si solar cell technology." *Photovoltaic Specialists Conference (PVSC), 2011 37th IEEE.* IEEE, 2011.

[27] Thouless M D, Evans A G, Ashby M F and Hutchinson J W 1987 Acta. Metall. 35 1333–41

[28] Suo Z and Hutchinson J W 1989 Int. J. Solids Struct. 25 1337–53

[29] Hutchinson J W and Suo Z 1992 Adv. Appl. Mech. 29 63–191

[30] Dross F, Robbelein J, Vandevelde B, Van Kerschaver E, Gordon I, Beaucarne G and Poortmans J 2007 Appl. Phys. A 89 149–52

[31] Thornton J A and Hoffman D W 1977 J. Vac. Sci. Technol. 14 164–8

[32] Dennis J K and Such T E 1993 Nickel and Chromium Plating 3rd edn (Cambridge: Woodhead)

[33] Murray C E, Saenger K L, Kalenci O, Polvino S M, Noyan I C, Lai B, and Cai Z 2008 J. Appl. Phys. 104 013530

[34] Shahrjerdi D et al 2012 Appl. Phys. Lett. 100 053901

[35] Hussain, Muhammad Mustafa, et al. "Gate-First Integration of Tunable Work Function Metal Gates of Different Thicknesses Into High-/Metal Gates CMOS FinFETs for Multi-Engineering." *Electron Devices, IEEE Transactions on* 57.3 (2010): 626-631.

[36] Sevilla, Galo A. Torres, et al. "Flexible and Transparent Silicon-on-Polymer Based Sub-20 nm Non-planar 3D FinFET for Brain-Architecture Inspired Computation." *Advanced Materials* 26.18 (2014): 2794-2799.

[37] Torres Sevilla, Galo A., et al. "Flexible Nanoscale High-Performance FinFETs."*ACS nano* 8.10 (2014): 9850-9856.

[38] Diab, Amer, et al. "Room to High Temperature Measurements of Flexible SOI FinFETs With Sub-20-nm Fins." (2014).

[39] J.-P. Colinge, FinFETs and Other Multi-Gate Transistors. New York, NY, USA: Springer-Verlag, 2007.

[40] C. Auth et al., "A 22 nm high performance and low-power CMOS technology featuring fully-depleted tri-gate transistors, self-aligned contacts and high density MIM capacitors," in Proc. Symp. VLSI Technol. (VLSIT), Jun. 2012, pp. 131–132.

[41] H. Xuejue et al., "Sub-50 nm p-channel FinFET," IEEE Trans. Electron Devices, vol. 48, no. 5, pp. 880–886, May 2001.

[42] J.-P. Colinge, "Multiple-gate SOI MOSFETs," Solid-State Electron., vol. 48, no. 6, pp. 897–905, 2004.

[43] D. K. Schroder, Semiconductor Material and Device Characterization, Wiley-IEEE Press , 2006

[44] Choi, Yang-Kyu, et al. "FinFET process refinements for improved mobility and gate work function engineering." *Electron Devices Meeting, 2002. IEDM'02. International.* IEEE, 2002.

[45] K. Akarvardar et al., "High-temperature performance of state-ofthe-art triple-gate transistors," Microelectron. Rel., vol. 47, no. 12, pp. 2065–2069, 2007.

[46] D. S. Jeon and D. E. Burk, "MOSFET electron inversion layer mobilities-a physically based semi-empirical model for a wide temperature range," IEEE Trans. Electron Devices, vol. 36, no. 8, pp. 1456–1463, Aug. 1989.

[47] J. Colinge et al., "Temperature effects on trigate SOI MOSFETs," IEEE Electron Device Lett., vol. 27, no. 3, pp. 172–174, Mar. 2006.

[48] J. Pretet, A. Vandooren, and S. Cristoloveanu, "Temperature operation of FDSOI devices with metal gate (TaSiN) and high-k dielectric," in Proc. 33rd Conf. Eur. Solid-State Device Res. (ESSDERC), Sep. 2003, pp. 573–576.

[49] D.-S. Jeon and D. E. Burk, "A temperature-dependent SOI MOSFET model for high-temperature application (27 °C–300 °C)," IEEE Trans. Electron Devices, vol. 38, no. 9, pp. 2101–2111, Sep. 1991.

[50] J. Wan, C. Le Royer, A. Zaslavsky, and S. Cristoloveanu, "Gate-induced drain leakage in FD-SOI devices: What the TFET teaches us about the MOSFET," Microelectron. Eng., vol. 88, no. 7, pp. 1301–1304, 2011.

[51] G. Ghibaudo, "New method for the extraction of MOSFET parameters," Electron. Lett., vol. 24, no. 9, pp. 543–545, 1988.

[52] S. M. Sze, Physics of Semiconductor Devices, 2nd ed. New York, NY, USA: Wiley, 1981

[53] G. Reichert, T. Ouisse, J. L. Pelloie, and S. Cristoloveanu, "Mobility modeling of SOI MOSFETs in the high temperature range," Solid-State Electron., vol. 39, no. 9, pp. 1347–1352, 1996.

[54] H.-C. Yuan, Z. Ma, M. M. Roberts, D. E. Savage, M. G. Lagally, J. Appl. Phys. 2006, 100, 013 708. b) H. Y. Li, L. H. Guo, W. Y. Loh, L. K. Bera, Q. X. Zhang, N. Hwang, E. B. Liao, K. W. Teoh, H. M. Chua, Z. X. Shen, C. K. Cheng, G. Q. Li, D.-L. Kwong, IEEE Electron Device Lett. 2006, 27, 538.

# CHAPTER SIX

## Inkjet-Printing of Conductive Tracks on Non-Woven Flexible Textile Fabrics for Wearable Applications

Kalyan Yoti Mitra[1], Dana Weise[1], Monique Helmert[1], Arvind Gurusekaran[1], Reinhard R. Baumann [1, 2]

[1]*Department of Digital Printing and Imaging Technology, Technische Universität Chemnitz, 09126 Chemnitz, Germany*

[2]*Department of Printed Functionalities, Fraunhofer Institute for Electronic Nano Systems ENAS, 09126 Chemnitz, Germany*

## Abstract

Inkjet printing is an attractive digital technology through which various functional materials like conductive inks can be deposited with a high level of accuracy in the micrometer range. Using this technology, various wearable electronic applications can be realized with a significant throughput of a small batch size having more customizable forms. In this chapter, inkjet printing of conductive tracks or electrode patterns on textile substrates is of the main focus. Three conductive inks are printed on three different flexible absorbing non-woven textile substrates. In this case, managing the printing process is of great interest and yet a great challenge, accounting the potential property of the conductive ink to flow in random directions into the textile substrate. Therefore, a morphological strategy is adopted and the electrical properties are investigated according to the spreading and blotting nature of the conductive inks into the textile, for developing the printed conductive tracks. The electrical characteristics of the conductive tracks are improved by the framed experimental set-ups in which the deposition parameters and also the post treatment process are subsequently varied. Furthermore, the post-treatment for the printed conductive tracks are up-scaled from sheet-to-sheet (S2S) which is via a conventional hotplate to Roll-to-Roll (R2R) manufacturing process by employing a free-lancing modular curing technology like Intense Pulsed Light (IPL) sintering.

**Keywords:** Inkjet printing, printed electronics, inkjet-printed conductive electrodes, non-woven textiles.

# 1. Introduction

Electrical conductive patterns e.g. electrodes on textiles are very common in medical devices for health monitoring like electrocardiogram (ECG) [1, 2]. The electrodes are fabricated by weaving [3], stitching [4] or knitting [2, 4] an electrically conductive thread onto the textile surface. Most of these stitched areas are more rigid than the rest of the textile and restrict a free mobility. Other articles investigate the production by printing technologies like screen printing [5, 6, 7], stencil printing [8] and inkjet printing [9, 10].

Here, especially the inkjet printing technology is challenging due to the low viscosity and the high solvent content of inkjet inks. Textiles have irregular surfaces with an absorbing characteristic, which promotes the ink to absorb and spread into the fiber developed where a hydrophobic barrier channel is first inkjet-printed to leave a reference for the succeeding conductive layer.

If this spreading cannot be controlled, patterns with defined dimension lose their shape and therefore lose or change their functionality. Another challenge is to form the functional conductive layers out of the printed non-conductive layers. Therefore, a sintering step is required, to evaporate all solvents, cast out the organic materials and merge the nanoparticles together [11, 12, 13]. For this sintering step, next to the conventional oven or hotplate sintering, various novel methods are known, e.g. chemical sintering [11], electrical sintering [11], plasma sintering [11], Infra-Red (IR) sintering [11, 14, 15] and intense pulsed light sintering (IPL) [11, 16, 17, 18]. These articles mainly focus on the sintering of various printed layers on non-absorbing substrates like polymeric foils or glass.

This chapter, on the other hand, demonstrates the inkjet printing of various conductive layers on a non-woven textile substrate. It focuses on the conventional hotplate sintering and the novel method of IPL sintering. Three different kinds of conductive inks have been used: one polymer based ink poly (3,4-ethylenedioxythiophene) polystyrene sulfonate, which is known as PEDOT: PSS and two metal-based nanoparticle inks-copper oxide (CuO) and silver (Ag), respectively. Their spreading and layer formation onto as well as into the absorbing textile substrate will be analyzed. Due to the different ink properties, the elected inks differ noticeably from each other not only concerning their flowing and penetration behavior but also on the electrical performance. By adjusting the printing parameters and optimizing the post treatment process, defined patterns with line resistance down to a few Ohms on the absorbing textiles could be achieved for the IPL sintering method.

## 2. Methodology

### 2.1 Materials and Deposition

Inkjet printing of the patterns was conducted using a Dimatix Materials Printer (DMP) 2831 from Fujifilm Dimatix and development material cartridge (DMC standard type) having nominal drop volume of 10 pL. Although the printing of all the conductive inks was performed using a drop space of 15 μm, the number of layers were varied from 1 to 3. The three non-woven textiles were chosen from Sächsisches Textilforschungsinstitut e.V. (STFI), Germany. Detailed information about these textiles are listed in Table 1. Also, a preliminary drop casting method was used to exploit the properties of the textile in the spreading nature of the different inks (microliter dispenser).

Table 1

| Textiles, Material | Tensile Force [N] | | Thickness [mm] | Porosity [l/m²/s] |
|---|---|---|---|---|
| | Direction | | | |
| | *Along Machine* | *Across Machine* | | |
| Textile 1: | Polypropylene / Polyethylene spun fabric (calendared) | | | |
| | 102.76 | 60.30 | 0.16 | 82.25 |
| Textile 2: | Polyethylene terephthalate/ Polyethylene spun fabric (calendared) | | | |
| | > 500 | > 500 | 0.18 | 59.67 |
| Textile 3: | Cellulose Wet fabric | | | |
| | 78.60 | 35.33 | 0.32 | 47.94 |

The conductive inks used here for developing the conductive patterns are polymeric and metal nano-particle based. Firstly, PEDOT: PSS water based ORGACONTM IJ-1005 conductive ink from Agfa-Gevaert N.V. was chosen. The ink has the following properties: solid content 0.8 wt %, viscosity 7 - 12 mPas, surface tension 31 - 34 mN/m. Secondly, two nano-particle based conductive inks were chosen for comparison. The first ink is Silverjet DGP 40LT-15C, Triethylene glycol monoethyl ether based Ag conductive ink from Advanced Nano Products which has the following properties: solid content 30 - 35 wt %, viscosity 10 - 17 cPs, surface tension 35 - 38 mN/m. The second ink is CuO-water based Metalon ICI-002HV conductive ink from Novacentrix, having the following properties: solid content ~ 12 wt %, viscosity $1 - 6.5$ cPs, surface tension ~ 35 mN/m. CuO ink using suitable post-treatment process is typically expected to reduce to Copper (Cu) as the conductive metal. Lastly, a UV curable polymer based dielectric ink Hyperion Pro Wet Black from Tritron GmbH is used for developing barrier channel, having 99 % polymer content. Details about the patterns of the barrier channel and conductive track for the entire electrode is shown in Fig. 1.

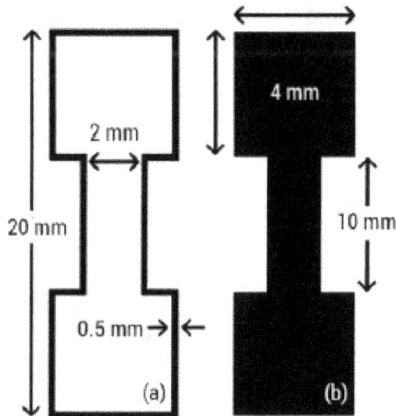

Figure 1 : Digital pattern of (a) UV barrier channel and (b) conductive track.

## 3. Curing and Characterization Methods

The entire fabrication process is explained in Fig. 2. Using the patterns and printing parameters described in the previous section, printing of the functional inks were performed. Every time the UV ink is kept the same for producing the barrier channel, but the conductive inks are varied. As can be observed from the first pattern shown in Fig 1 (a) that it is printed via varying the number of layers followed by UV curing for 20 sec with a substrate distance of 10 cm (indicated with purple flash sign); this concludes step no. 1. Next is printing of the second pattern as shown in Fig. 1 (b) with the conductive inks separately but with varying the number of layers followed by thermal curing (indicated with red flash sign). The intention here is to control the spreading of the conductive inks and to achieve the functional electrode patterns.

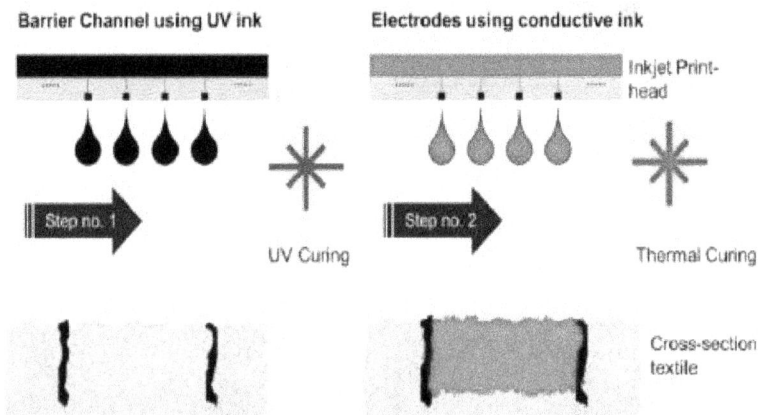

Figure 2: Fabrication schemes for developing conductive electrodes on textiles. Printing of a barrier channel as a first step and printing of the conductive ink inside the barrier as a second step.

The thermal sintering for the printed conductive layers on the textiles was conducted using the standard hotplate and the IPL sintering method. For IPL sintering, the PulseForge 3200 from Novacentrix was used.

The curing parameter for the hotplate was kept constant at 150 °C for 30 min for all the three conductive inks. The curing parameters for IPL were adjusted and varied in accordance to the ink used. The used parameters are listed below in Table 2, where each set of parameters is categorized as setting 1 to 8.

Table 2

| Setting number | 1 | 2 | 3 | 4 | 5 | 6 | 7 | 8 |
|---|---|---|---|---|---|---|---|---|
| Pulse duratio n [ms] | 1 | 4 | 6 | 3 | 6 | 4 | 6 | 6 |
| Energy density [J/cm$^2$] | 1.41 | 1.98 | 2.66 | 3.52 | 4.04 | 4.32 | 5.57 | 6.62 |

Characterization of the textiles was performed optically using a light microscope and light transmittance mode. Once the printing and curing of the functional material inks are accomplished, the electrical characterization is conducted by measuring the resistance using a standard electronic LCR multi-meter optimized at a distance of 1.5 cm across the contact pads of the printed electrodes.

## 4. Results and Discussion

### 4.1 Optical Characterization

The images of the three textiles along with the corresponding microscopic images are depicted in Fig.3. It can be seen that the fiber orientation of the three textiles are greatly varying. Textile 1 and 2 show loosely packed fibers. On the contrary, Textile 3 is denser which can be attributed to a larger amount of the material per unit area. The fiber types also vary due to the constituting material itself. These properties can affect the spreading behavior of the ink directly.

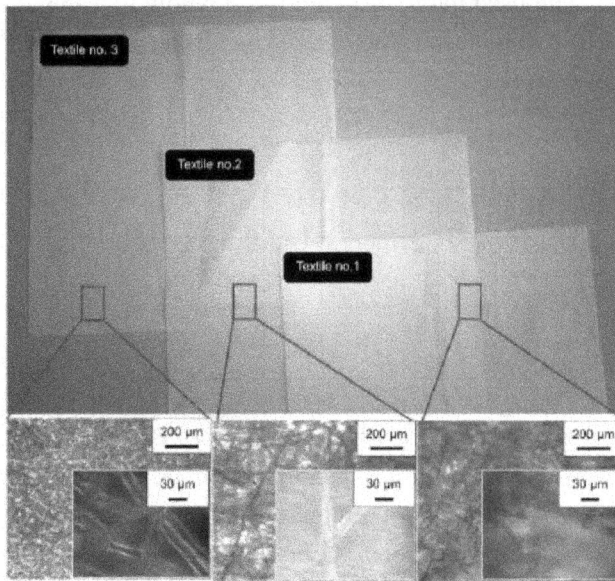

Figure 3: Photograph and microscopic images (transmission light) for textile 1, 2, and 3, respectively.

In Fig. 4 (a), images are shown for a dispensed single droplet using a microliter dispenser (volume/drop: 10 μL) of the UV curable ink on the three different textiles. By inspecting the image, it can be concluded that spreading of the UV ink is minimum for Textile 3 and maximum for Textile 1. Therefore,

Textile 3 was chosen as the candidate substrate for further printing investigation. Next, and as depicted in Fig. 4 (b), the drop spreading is explored when the same volume of the drop is dispersed with different functional inks on the same Textile. One can easily distinguish that spreading of the CuO ink is maximum, and somehow similar for the PEDOT: PSS and Ag inks. This can be attributed to the rheological properties of the inks. It is also believed that the solvent in CuO ink can accelerate the capillary force within the non-woven textile substrate leading to a higher blotting phenomenon.

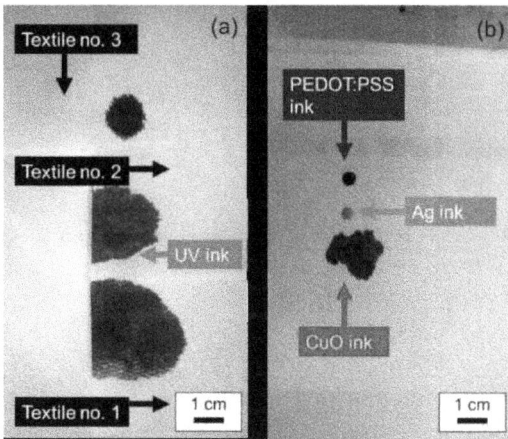

Figure 4: Images showing different spreading of the used inks on the used textile substrates: (a) UV ink drop casted on the three different substrates; (b) three different functional inks drop casted on textile 3.

It is clearly visible from Fig.5 that when the number of printed layers are increased, the width of the barrier channel also increases. It is also expected that when the number of the layers is increased the expected strength of the pattern to withhold the conductive ink inside also improves. Considering the drop cast PEDOT: PSS to the one, two and three layers of barrier channels, the flow of the ink is controlled easily as shown in Fig. 5 (d). On the other hand, when Ag ink is considered, the ink is not easily controlled as shown in Fig. 5 (c). Also, when CuO ink is considered, three layers of the barrier channel are required at the minimum to restrict the ink from flowing out of the barrier channel.

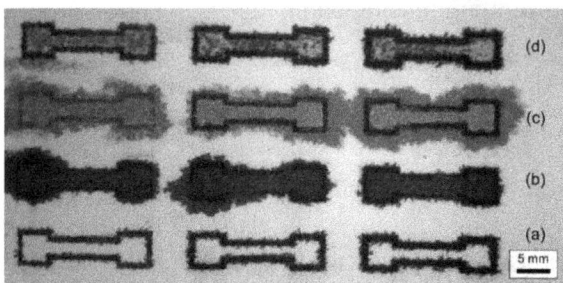

Figure 5: Images of patterns on Textile 3: (a) inkjet-printed UV ink barrier channels; (b) drop casted CuO, (c) drop casted Ag, (d) drop casted PEDOT: PSS.

Figs. 6 (a) - (c) illustrate the barrier channel printed with three layers on Textile 3; and PEDOT: PSS, Ag and CuO inks also printed within the previously printed barrier channel. It should be noted that minimum three layers of PEDOT: PSS, two layers of Ag, and two layers of CuO inks were found to be the most optimal for developing such conductive tracks on Textile 3. These were found to be the optimal printing parameters as to achieve the conductive tracks on Textile 3. The ink is clearly seen to be restrained by the barrier channel. Also, when compared to the drop casted ones mentioned before, the flow restriction of the conductive inks was far easier due to the existence of the printing and drying phenomena simultaneously, and to the reduced volume of deposition.

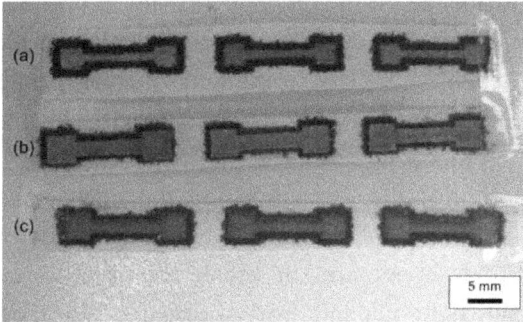

Fig. 6. Images of three inkjet-printed UV barrier channel layers with inkjet-printed (a) three layers of PEDOT:PSS, (b) two layers Ag, and (c) two layers CuO ink on Textile 3

In this section, the blocking ability of the number of barrier channel layers are shown as a function of the drop casted and inkjet-printed ink layers on Textile 3.

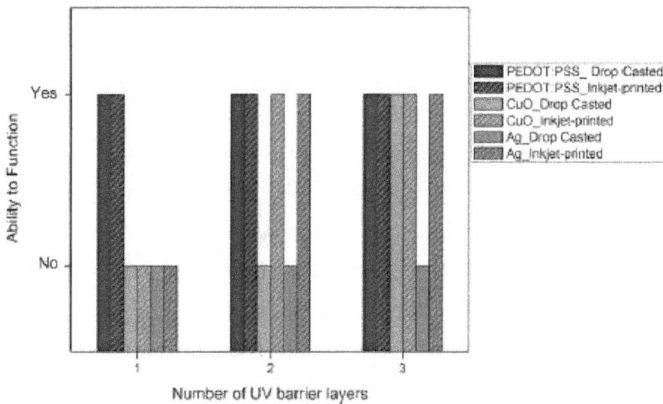

Figure 7: Graph showing the strength of the barrier channel as function of drop casted and inkjet-printed conductive inks on Textile 3.

It is worth mentioning that when PEDOT: PSS is inkjet-printed (three printed layers) or even drop casted, it does not make a significant influence on the outward flowing properties because, in both conditions, the deposited layer is well restricted within the barrier channel with a minimum of one printed barrier channel layer. When CuO ink is considered with drop casting deposition method, three printed barrier channel layers are required to control the spreading. On the other hand, if CuO inkjet printing (two printed layers) is considered then two printed barrier channel layers are, at most, required to control the spreading. When Ag ink and drop casting deposition are together considered, even three barrier channel layers were found to be incapable of controlling the spreading of the ink. Although, when the amount of deposited Ag is reduced via inkjet printing (two printed layers) and the drying process is accelerated, then the blocking ability to control the spreading becomes higher.

## 4.2 Electrical Characterization

This section discusses the results of the electrical resistance which the inkjet-printed conductive tracks exhibit at different curing conditions. The considered conditions are: hotplate and IPL for PEDOT: PSS (three printed layers), Ag (two printed layers), and CuO (two printed layers) ink. Two layers for Ag and CuO are expected to be optimal, due to the insignificant reduction of the electrical resistance when both 2 and layers are compared. The fabricated conductive electrodes were developed using three barrier channel layers on Textile 3. Every time, five samples are taken into account for the measurement process. The electrical resistance from the fabricated electrodes were measured using multi-meter with the measurement probes kept at 1.5 cm.

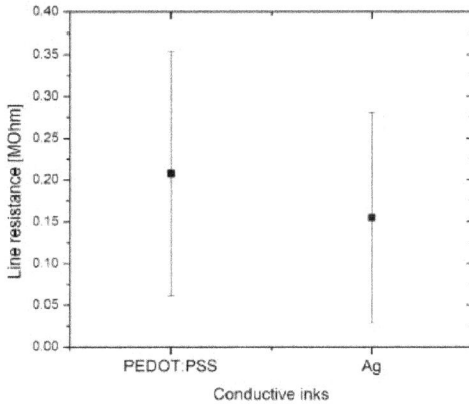

Figure 8: Graph showing electrical resistance for inkjet-printed PEDOT: PSS (three layers) and Ag (two layers) on Textile 3 with three barrier channel layers cured on hotplate

It is assumed that the curing process using the hotplate was not penetrative enough to accomplish lower resistance values. On the other hand due to the requirement of high temperature and energy for the chemical reduction of CuO to Cu, the ink showed no real resistance even in the $M\Omega$ region. Thus, it can also be considered as non-conductive or highly resistive.

As an alternative to the conventional hotplate curing process, as mentioned in the experimental section, IPL curing was chosen as a secondary sintering methodology. All the curing parameters described in Table 2 were implemented for the identically fabricated conductive electrodes of PEDOT: PSS (three printed layers), Ag (two printed layers), and CuO (two printed layers) inks mentioned before and cured using the hotplate. But, the number of UV barrier channel layers were varied from one, two, and three.

Considering curing of three PEDOT: PSS layers, IPL energy densities ranging from 1.4 to 4.3 J/cm2 resulted in relatively higher resistance in the MΩ, whereas with 5.571 J/cm2, the resistance was brought down considerably to kΩ. Furthermore, an increment in the implementation energy density leads to no significant variation.

In Fig. 9, a graph is depicted where the IPL curing parameter is kept constant at 5.6 J/cm2 and the function of electrical resistance is shown against the number of barrier channel layers. As the number of barrier channel layers is increased, the resistance of the cured PEDOT: PSS layers decreases, ranging between 7 – 22 kΩ. This dependency of the resistance on the barrier channel layers infers the reduction of the area where the conductive material is deposited. In such situations, the PEDOT: PSS ink tends to provide its movement in the z-axis direction which in this case is the depth of Textile 3. Thus, the dependence is proportional.

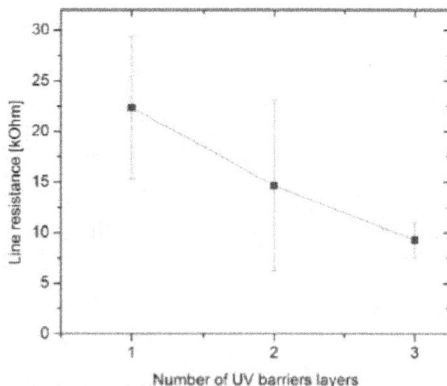

Figure 9: Graph showing electrical resistance for three inkjet-printed PEDOT: PSS layers as a function of UV barrier layers cured with IPL at energy density of 5.6 J/cm2

The previous process is followed by depositing the CuO ink layer, where the electrical resistance was very high, as noted earlier, via hotplate sintering. For the CuO ink layers, a similar range of energy density was varied. With the IPL energy densities ranging from 1.4 to 5.6 J/cm2 no noticeable electrical resistance in MΩ regime could be achieved. Whereas with 6.6 J/cm2, the resistance was brought down to MΩ. Typically the value ranged between 2 - 3 MΩ. A graph depicted in Fig. 10 shows the IPL curing parameter where it is kept constant at 6.6 J/cm2 and the function of electrical resistance is shown against the number of barrier channel layers.

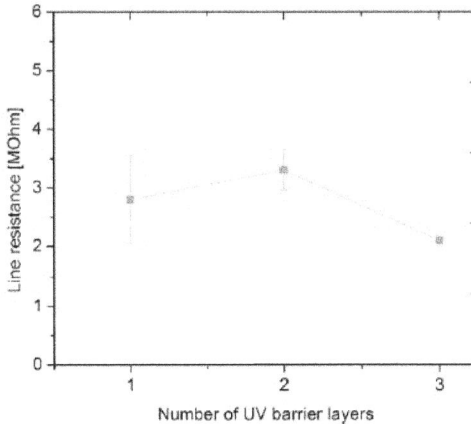

Figure 10: Graph showing electrical resistance for two inkjet-printed CuO layers as a function of UV barrier layers.

Lastly, the Ag ink is evaluated, which proves to be the best conductive ink in terms of curing via conventional hotplate, thereby showing the least electrical resistance. Next, the electrical characteristics of the conductive tracks are compared with IPL to that of the hotplate curing. As mentioned previously, the curing energy density was varied between 1.41 - 6.623 J/cm2. It was observed that the impact of curing towards Ag layers starting from the lowest energy density resulted in lowering the electrical resistance. As shown in Fig. 11, a resistance as low as ~ 4 Ω could be obtained when the energy density of IPL curing is tuned at 4.32 J/cm2. It should be noted that any further increment in the energy density would directly burn Textile 3 and destroy the Ag layer completely. As noticed in the previous case, PEDOT: PSS and CuO ink also exhibit an increment in the number of UV barrier channels which is associated with a decrement in the resistance of the fabricated electrodes.

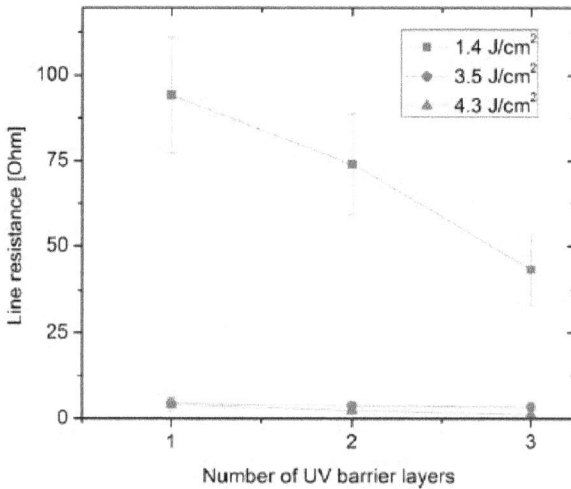

Figure 11: Graph showing electrical resistance for two inkjet-printed Ag layers as a function of UV barrier layers cured with IPL at different energy densities

## 5. Conclusion

In this chapter, we have shown a two-step fabrication process for manufacturing inkjet-printed conductive electrodes on textile substrates using UV barrier channels and low viscosity functional inks. A direct dependence of the resistance of the fabricated electrodes on the strength of the barrier channel is observed. This paves the way towards the manufacturing of electrodes ($< 5$ $\Omega$/cm) on textiles substrate, thereby, allowing more customization via digital printing technology of wearable electronics that can be integrated and utilized for different applications. In general, it was shown that there are various options for the choice of conductive inks and the required range of resistance. Ag and PEDOT: PSS inks are better choices compared to CuO due to the lower resistance values. In all cases, manufacturing of the electrodes can be achieved through an R2R processing platform.

### Acknowledgement

The authors would like to thank the Sächsisches Textilforschungsinstitut e.V. (STFI) for providing the textile samples. We also thank the Fraunhofer ENAS institute, department of printed functionalities, for using their characterization equipment. The authors are also grateful for the financial support from the funding sources: 1) EU funded project: TDK4PE, grant agreement no.: 287682, 2) Deutsche Forschungemeinschaft (DFG) funded project: Federal Cluster of Excellence "MERGE Technologies for Multifunctional Lightweight Structures", grant agreement no.: EXC 1075, and 3) DFG funded project: MIKROHIPS, grant agreement no.: FKZ: BA 1169/7-1.

# References

[1] T. Poly, J. Vanhala, "Textile Electrodes in ECG Measurement", ISSNIP 2007, Melbourne, Qld., 2007, pp. 635 - 639.

[2] L. Vojtech, R. Bortel, M. Neruda, M. Kozak, "Wearable Textile Electrodes for ECG Measurements", Biomedical Engineering, vol. 11, pp. 410 - 414, 2013.

[3] A. Dhawan, A. M. Seyam, T. K. Ghosh, J. F. Muth, "Woven Fabric-Based Electrical Circuits-Part I: Evaluating Interconnect Methods", Textile Research Journal, 74 (10), pp. 913 - 919, 2004.

[4] L. Li, W. M. AU, K. M. Wan, S. H. Wan, W, Y, Chung, K. S. Wong, "A Resistive Network Model for conductive Knitting Stitches", Textile Research Journal, vol. 80 (10), pp. 935 - 947, 2010.

[5] I. Kazani, C. Hertleer, G. De Mey, A. Schwarz, G. Guxho, L. Van Langenhoven, "Electrical Conductive Textiles Obtained by Screen Printing", FIBRES & TEXTILES in Eastern Europe, 20, Vol. 1 (90), pp. 57 - 63, 2012.

[6] G. Paul, R. Torah, S. Beeby, J. Tudor, "Novel active electrodes for ECG monitoring on woven textiles fabricated by screen and stencil printing" SENSORS AND ACTUATORS A-PHYSICAL, Vol. 221, pp. 60 - 66, 2015.

[7] G. Paul, R. Torah, K. Yang, S. Beeby, J. Tudor, "An investigation into the durability of screen-printed conductive tracks on textiles", MEASUREMENT SCIENCE & TECHNOLOGY, Vol. 25 (2), Article Nr. 025006, 2014.

[8] N. Matsuhisa, M. Kaltenbrunner, T. Yokota, H. Jinno, K. Kuribara, T. Sekitani, T. Someya, "Printable elastic conductors with a high conductivity for electronic textile applications", NATURE COMMUNICATIONS, Vol. 6, Article Nr. 7461, pp. 1 - 11, 2015.

[9] Z. Stempien, E. Rybicki, T. Rybicki, J. Lesnikowski, "Inkjet-printing deposition of silver electro-conductive layers on textile substrates at low sintering temperature by using an aqueous silver ions-containing ink for textronic applications", SENSORS AND ACTUATORS B-CHEMICAL, Vol. 224, pp. 714 - 725, 2016.

[10] H. F. Castro, E. Sowade, J. G. Rocha, P. Alpuim, A. V. Machado, R. R. Baumann, S. Lanceros-Mendez, "Degradation of all-inkjet-printed organic thin-film transistors with TIPS-pentacene under processes applied in textile manufacturing", ORGANIC ELECTRONICS, Vol. 22, pp. 12 - 19, 2015.

[11] A. Kamyshny, J. Steinke, S. Magdassi, "Metal-based Inkjet Inks for Printed Electronics", The Open Applied Physics Journal, Vol. 4, pp. 19 - 36, 2011.

[12] J. R. Greer, R. A. Street, "Thermal cure effects on electrical performance of nanoparticle silver inks", Acta Materialia, Vol. 55, pp. 6345 - 6349, 2007.

[13] D. J. Lee, J. H. Oh, H. S. Bae, "Crack formation and substrate effects on electrical resistivity of inkjet-printed Ag lines", Materials Letters, Vol. 64, pp. 1069 - 1072, 2010.

[14] E. Sowade, H. Kang, K. Y. Mitra, O. J. Weiß, J. Weber, R. R. Baumann, "Roll-to-roll infrared (IR) drying and sintering of an inkjet-printed silver nanoparticle ink with 1 second", Journal of Material Chemistry, vol. 3, pp. 11815 - 11826, 2015.

[15] Daniel Tobjörk, Harri Aarnio, Petri Pulkkinen, Roger Bollström, Anni Määttänen, Petri Ihalainen, Tapio Mäkelä, Jouko Peltonen, Martti Toivakka, Heikki Tenhu, Ronald Österbacka, "IR-sintering of ink-jet printed metal-nanoparticles on paper", THIN SOLID FILMS, vol. 520, no. 7, pp. 2949-2955, 2012.

[16] D. Weise, R. R. Baumann, "Intense pulsed light sintering and parameter optimization of various inkjet printed silver electrodes", Digital Fabrication and Digital Printing, NIP31, Portland, Oregon, Vol. 2015, Number 1, pp. 62-65 (4), 2015.

[17] J. Perelaer, R. Abbel, S. Wünscher, R. Jani, T. van Lammeren, U. S. Schubert, "Roll-to-Roll Compatible Sintering of Inkjet Printed Features by Photonic and Microwave Exposure: From Non-Conductive Ink to 40% Bulk Silver Conductivity in Less Than 15 Seconds", ADVANCED MATERIALS, Vol. 24 (19), pp. 2620 - 2625, 2012.

[18] J. Niittynen, R. Abbel, M. Mäntysalo, J. Perelaer, U. S. Schubert, D. Lupo, "Alternative sintering methods compared to conventional

thermal sintering for inkjet printed silver nanoparticle ink", THIN SOLID FILMS, Vol. 556, pp. 452 - 459, 2014.

# CHAPTER SEVEN

## Wearable Electronics for Intra-Body Communications

Asimina Kiourti

ElectroScience Laboratory, Dept. of Electrical and Computer Engineering, The Ohio State University, Columbus, Ohio, USA.

## Abstract

Rapid advances in wireless communications, electronics, sensing technologies, and materials are opening new and hitherto unexplored opportunities. One of the latest developments is in the area of wearable electronics, i.e., devices that are integrated into clothing and/or accessories. Wearable electronics are recently taking off for a number of applications, including healthcare, sports, defense, etc. In the most common scenario, the wearable device consists of electronics and an antenna. The latter is used to wirelessly transmit information between the wearable device and a cell phone, tablet, or other Personal Digital Assistant (PDA) device. Challenges in this case include signal interference from other concurrent wireless transmissions, as well as high power consumption. To address these challenges, a novel category of wearable devices has recently started to emerge, that of intra-body communicating wearables. Intra-body communications (IBC) use the human body as a communication medium. That is, devices that are placed on or near the human body can transmit/receive information through the biological tissues. This chapter provides an overview of existing wearable technologies for IBC, addresses their potential and challenges raised, and discusses future directions.

## 1. Introduction

We live in the era of wireless communications 0 0. Nowadays, wireless devices are part of everyone's daily lives, while novel wireless devices are constantly emerging, aiming to improve our quality of life. Some typical examples of widely used wireless devices include cell phones, tablets, laptops, wireless keyboards, Wi-Fi routers, wireless headsets, etc.

Recently, a new trend is taking off that aims at integrating wireless and smart capabilities to our traditional clothing and accessories (e.g., shirts, socks, bracelets, glasses, etc.) 0. This is done by unobtrusively integrating miniature wireless sensors within the clothing and/or accessories. These sensors perform a number of smart functionalities (e.g., sensing our body temperature, heart rate, number of steps, etc.), and eventually transmit the collected information to a central node that is placed further away. The latter is typically a Personal Digital Assistant (PDA) device, such as a cell phone, tablet or laptop.

Several wearable devices with wireless functionalities are already available in the market. Some examples are: (a) smart watches that include heart rate sensors, global positioning systems (GPS), and accelerometers to measure "the many ways we move" 0, (b) biometric/fitness T-shirts that monitor heart rate, breathing rate, activity intensity, sleep patterns, etc. 0, 0, (c) easy to wear T-shirts that enable physicians to do assessment of cardiac problems in an everyday environment 0, (d) smart socks that detect parameters important to the running form, including cadence and foot landing technique 0, and (e) activity tracker bracelets that monitor activity, sleep stages, calories, heart rate, etc. 0. All these wearable devices work in a very similar way. As a first step, the on-body sensor detects the desired physiological parameter (e.g., heart rate, acceleration, etc.). Further, a wireless module with an integrated antenna transmits this collected information to a mobile device that is placed further away. Wireless transmission may be performed through typical wireless communication protocols such as Bluetooth, Zigbee, or Wi-Fi, among others. Eventually, the collected data are displayed using an app on the individual's mobile device.

The aforementioned wireless communication approach is very promising, and intensive research is currently being performed to further improve its quality and performance 0-0. However, the concept of wireless data transmission from a wearable device to a PDA device, using air as the transmission medium, has a number of challenges. These include: (a) signal interference from other wireless transmissions that are taking place in the same environment at the same time, and (b) high power consumption by the wearable device implying a need for batteries that require frequent replacement and/or recharging. To address these challenges, a novel category of wearable devices has recently started to emerge, that of intra-body communicating wearable devices.

Intra-body communications (IBC) use the human body as a communication medium 0. That is, devices that are placed on or near the human body can transmit/receive information through the conductive biological tissues. Advantages in this case include minimal to zero signal interference, and very low power consumption by the wearable device(s). This chapter provides an overview of existing wearable technologies for IBC, addresses their potential and challenges raised, and discusses applications and future directions.

## 2. What are Intra-Body Communications?

The concept of IBCs was originally proposed by IBM and Zimmerman in 1996 0. In that original work, Zimmerman demonstrated how mobile devices placed upon or near the human body could exchange data and other digital information through the human body itself. Zimmerman's original prototype transmitter operated at 330 kHz, and had a transmission power consumption of 1.5 mW.

Nowadays, IBCs are becoming more and more popular. In principle, IBCs use the human body, and therefore the conductive biological tissues, as a signal transmission medium. In other words, some part of the signal propagates through the human body tissues rather than through the air. In the simplest scenario, two wearable devices are communicating through IBCs. In a more complicated scenario, several wearable devices exist upon the body. These devices (transmitters) can be designed to send the collected information to a single on-body receiver, using IBCs. The receiver device itself may further transmit all gathered information to a distant PDA. The latter communication is typically performed using a standard wireless protocol, such as Bluetooth, Zigbee, or Wi-Fi.

Depending on how the electrical signals are transmitted from the transmitter to the receiver using the human body as a transmission medium, IBCs can be divided into two major categories: (a) galvanic coupling, and (b) capacitive coupling 0, 0. Both of these approaches rely on low-frequency and low-level currents/voltages through the human body. Different researchers have used different frequencies and current levels. However, typically, IBC frequencies are below 1 MHz, and the peak current is lower than 2 mA. For galvanic coupling IBCs, the signal is controlled by a current flow through the body.

For capacitive coupling IBCs, the signal is controlled by an electric potential between the transmitter electrodes. Further details regarding these two IBC approaches are discussed next.

### a)  Galvanic Coupling

The concept of galvanic coupling is illustrated in Fig. 1 0. As mentioned above, for galvanic coupling IBCs, the signal is controlled by a current flow through the body. Specifically, and as seen in Fig. 1, two pairs of electrodes are attached to the skin, the transmitter pair and the receiver pair. The transmitter pair of electrodes is used to inject an electric current into the body. As would be expected, most of the current flows through the direct path between the transmitter electrodes. However, since the underlying biological tissues are conductive, there are some weak currents that are propagating inside the body. Therefore, a small current travels from the transmitter electrode pair to the receiver electrode pair. The latter generates a voltage drop that can easily be detected by the receiver electronics. Using the galvanic approach, the signal propagates entirely through the human body. Therefore, the received signal is not affected by the surrounding environment in any way, and is not susceptible to any sort of exterior environmental interference. This is a major advantage of galvanic coupling IBCs.

Operation frequencies used for galvanic coupling are typically in the 10 kHz to 1 MHz range. Another consideration is the patient safety against electromagnetic fields. As is well-known, exposure to electromagnetic fields results in internal body currents and power absorption by the biological tissues. To avoid potential severe health effects, one must conform to restrictions set by the Federal Communications Commission (FCC), the Institute of Electrical and Electronics Engineers (IEEE), and the

International Commission on Non-Ionizing Protection (ICNIRP), among others. For example, ICNIRP 0 sets: (a) current density restrictions for frequencies from 1 Hz to 10 MHz, (b) specific energy absorption rate (SAR) restrictions in the frequency range from 100 kHz to10 GHz, and (c) power density restrictions in the frequency range from 10 GHz to 300 GHz. Current density restrictions in the frequency range that is of interest to IBCs are provided to prevent nerve stimulation, viz. adverse effects on the regular functions of the nervous system. The most stringent restrictions are set in the frequency range between 4 Hz and 1 kHz, where the maximum current density is set to 2 mA/m$^2$. To comply with the aforementioned patient safety requirements, injection currents for galvanic coupling IBCs are typically in the range of 1 mA to 2 mA.

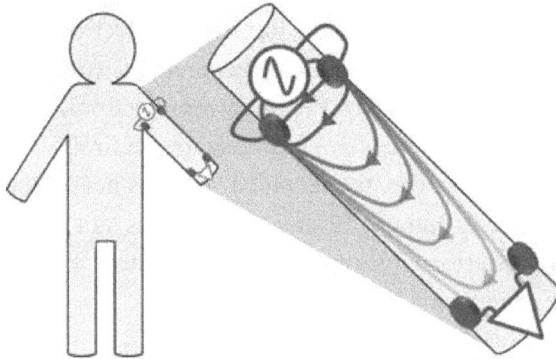

Figure 1: Operation principle of galvanic coupling IBCs 0.

### b)  Capacitive Coupling

The idea behind capacitive coupling is illustrated in Fig. 2 0. Same as for galvanic coupling, two pairs of electrodes are employed: the transmitter pair and the receiver pair. Each pair consists of a signal electrode and a ground electrode. The signal electrodes of both the transmitter and the receiver are attached to the human body. Instead, the ground electrodes are detached from the human body, or "floating". As mentioned above, for capacitive coupling IBCs, the signal is controlled by an electric potential between the transmitter electrodes. Specifically, for IBC to occur, an electric field is induced between the transmitter electrodes (signal and ground). Part of this field travels from the transmitter's signal electrode through the underlying biological tissues and eventually through the air or the external ground in order to return back to the transmitter's ground electrode. Concurrently, a small part of this field flows through the receiver's signal and ground electrodes. Signal transmission is achieved by modulating the voltage at the transmitter electrodes. This modulated signal is eventually detected and decoded by the receiver electronics. Based on the above, there are four components involved in capacitive coupling IBCs: (a) the human body, (b) the signal electrodes, (c) the ground electrodes, and (d) the external ground. Coupling between the electrodes and the human body, the air, and the external ground can be modeled as capacitances, hence the name "capacitive coupling".

Same as for galvanic coupling, capacitive coupling IBCs need to conform to patient safety requirements in order to prevent adverse health effects. Typically, the induced current conducted through the body is in the order of magnitude of picoamperes. As such, it is not harmful to the human body. Nevertheless, it is noted that different research groups have been using different configurations and different electrodes, implying that there is no "rule of thumb" regarding the optimum electric field amplitudes, or even frequencies, for IBCs. In general, however, frequencies used for capacitive coupling IBCs are usually higher than those used for galvanic coupling. It is also noted that the "floating" electrodes required in capacitive coupling make the implementation of capacitive coupling transmitters and receivers quite challenging.

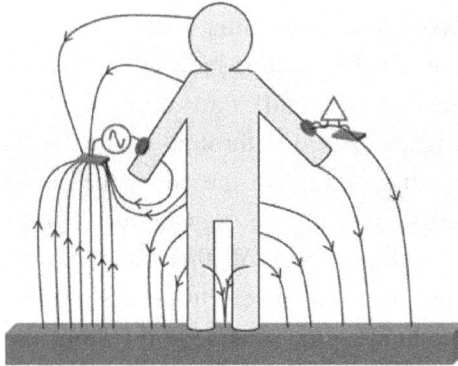

Figure 2: Operation principle of capacitive coupling IBCs 0.

## 3. Wearables for Intra-Body Communications

IBC transmitters and receivers need to be placed on or near the human body. Therefore, to ensure unobtrusiveness, such transmitters and/or receivers can be integrated with wearable devices, such as smart watches, bracelets, and other multimedia devices.

The main components of a typical IBC system are shown in Fig. 3 0. In general, the IBC transmitter device allows sensed data to be compressed and encoded, and eventually transmitted through a pair of electrodes. That is, the transmitter typically includes a sensor, analog-to-digital converter (ADC), encoder, modulator, and coupler. The signal carrier is modulated by modifying the amplitude (on-off keying: OOK), frequency (frequency shift keying: FSK) or phase of the carrier (phase shift keying: PSK).

The human body serves as a communication channel. Finally, the electrical signals that are coupled to the body are collected by the IBC receiver. The IBC receiver device consists of the receiving electrodes, an analog detector unit that amplifies the recorded signal, and digital components for data demodulation, decoding, and extraction. That is, the receiver typically includes a detector, demodulator, signal decoder, and digital-to-analog converter (DAC).

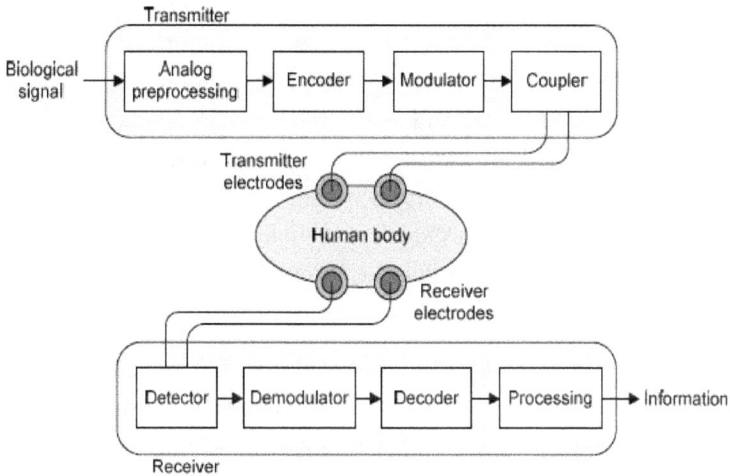

Figure 3: Block diagram of a typical IBC system 0.

As would be expected, a critical challenge in designing wearable IBC devices is to minimize their size, power consumption, hardware complexity, and cost. At the same time, IBC transceivers should be simple, unobtrusive to the wearer, lightweight, and flexible. Though energy consumption of IBC transceivers is small, there is still some energy needed to power them up. The simplest solution would be to integrate batteries.

However batteries require often replacement and/or recharging, and they are not as unobtrusive as someone would want them to be for wearable applications. Therefore, alternative power sources, such as solar cells or radio-frequency (RF) power harvesters may be employed instead.

In designing wearable IBC transceivers, it is also important to highlight that electrodes themselves are highly critical. This is because electrodes play a major role in ensuring good coupling with the human body. Geometry, surface area, the radius of curvature and even position of the electrodes on the human body can significantly affect the amount of signal coupled to the human body. Concurrently, electrodes should be easy to use, comfortable to wear, and not trigger any skin reactions. Typical electrodes used in proof-of-concept experimental set-ups are fabricated using commercial metals, such as aluminum, copper, bronze, brass, stainless steel, and nickel silver. In future, flexible electrodes made of conductive threads 0 0, and conductive inks 0 can be employed instead.

Overall, there is no "rule of thumb" for optimizing an IBC system design. It is up to the designer to come up with an optimal, per case and per application, design. As an example, Fig. 4 shows an IBC system using a 10 MHz carrier frequency 0. The user wears the transmitter electrode pair. When he/she touches the electrode of the receiver device, a transmission channel is formed through the human body. For this particular example, the receiver recognizes the user's identity ID.

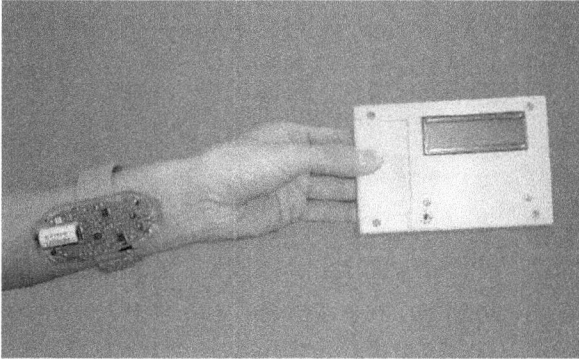

Figure 4: Example IBC system: for this particular example, the receiver recognizes the user's identity (ID) 0.

## 4. Performance of Intra-Body Communication Systems

As mentioned above, IBCs have lower power consumption and are more secure than common wireless communication technologies such as Bluetooth, ZigBee, and Wi-Fi. Also, since IBCs form a short-range communication network inside and around the human body, they allow the same frequency band to be reused by other users with minimal to zero interference.

Regarding power consumption, IBCs require very small power for transmission and reception. The characteristics of the human body favor operation at carrier frequencies between 100 kHz and 500 kHz. This frequency is much lower than that used by traditional wireless protocols such as Bluetooth or Wi-Fi (the latter carrier frequencies are in the GHz range). These low carrier frequencies imply low data bit rates and low complexity modulations. In doing so, they provide for very-low-power designs. Due to the reduced power consumption, heating and tissue irritation of the patients are lower, and the battery lifetime is longer than traditional wireless technologies (Bluetooth, ZigBee, and Wi-Fi). We note that energy consumed by IBC devices can be divided into three major categories: (a) power consumption of the sensor signal conditioning and storage units, (b) power consumption of the baseband signal processing, and (c) power dissipation of the analog front-end. A system analysis shows that the power consumption is mainly dominated by the system clock frequency of the digital units and the output power of the coupler units.

Regarding security, since signals propagate through the human body, electromagnetic noise and interference have little effect on data transmission. Compared to capacitive coupling IBCs, galvanic coupling IBCs are less prone to environment interference. This is because the galvanic coupling signal/ground electrodes are directly attached to the human body. In other words, the transmission path of galvanic coupling IBCs solely depends on the human body. Therefore, it does not interact with the surrounding environment in any way, implying much higher security for the transmitted data. In contrast, for capacitive coupling IBCs, the two ground electrodes require "floating". Therefore, in this case, the surrounding environment turns out to have a significant influence on the overall quality of communication.

Another advantage of galvanic coupling vs. capacitive coupling IBCs is the use of lower frequency carriers. The latter slows down the system clock and, therefore, simplifies transceiver design 0. As would be expected, this advantage of galvanic coupling IBCs comes at the expense of reduced data rates. Nevertheless, the achieved data rates are still adequate and acceptable for typical wearable sensing applications. For example, a glucose monitor requires data rates of around 1.6 kbps, a body temperature surveillance system required data rates of 120 bps, whereas ElectroCardioGram (ECG) signal collection requires data rates of 144 kbps 0. Of course, the characteristics of the human body channel play a significant role in the performance of IBC systems. The IBC channel that is formed by biological tissues works very much like a filter: it introduces loss and attenuates the signal. Specifically, the transmitter-receiver electrode distance affects signal attenuation, with larger distances resulting in higher signal attenuation. Furthermore, frequency-dependent gain characteristics and multipath effects distort the transmitted signal. Different IBC frequencies, different modulation methods, different electrode placement upon the human body, and even different human subjects imply different signal attenuations and, thus, different Bit Error Rates (BER).

Along these lines, an important issue to consider in assessing the performance of IBCs is the effect of body movement. Whole body motions, such as sitting, standing, and walking, have been shown to not significantly affect channel attenuation 0, 0. Instead, flexing of the forearm and elbows in a scenario of upper limb capacitive coupling IBCs, has shown a variation in channel gain by around 2-5 dB 0. In another study 0, results showed that the presence of limb joints causes high signal attenuation, up to 8 dB. Moreover, signal attenuation has been shown to decrease with decreasing joint angle. For example, in one case, flexing the elbow joint resulted in 5 dB of signal attenuation at 10 MHz. Importantly, results have indicated that IBCs are most likely to be influenced by movement at frequencies below 50 MHz. Therefore, adaptive power control is often suggested as a solution to address these channel variabilities.

Overall, this new wireless data wireless data communication technology of IBCs promises new and hitherto unexplored opportunities for a very wide range of applications. Several research groups are working in this area with the developed systems being different in terms of the coupling methods, induced current amplitudes, selected operation frequencies, signal modulation methods, and achieved data rates. A comparison of intra-body communication systems reported to date is shown in Table I 0-0 (adopted from 0). We note that there are two methods used to model and simulate an IBC system, and different research groups may employ either of these methods. Specifically, the first method develops a transfer function to describe the mathematical relations of the different parts 0, 0. The second method develops a finite-element model of the human body parts 0, 0.

Table 1: Comparison of intra-body communication systems 0.

| Ref. | Year | Coupling method | Coupling amplitude | Carrier frequency | Modulation | Data rate |
|------|------|-----------------|--------------------|--------------------|------------|-----------|
| 0 | 1995 | Capacitive | 30 V | 330 kHz | OOK | 2.4 kbps |
| 0 | 1997 | Capacitive | 21 V | 90 kHz | FM | 0.1 kbps |
| 0 | 2001 | Capacitive | 22 V | 160 kHz | FSK | 38.4 kbps |
| 0 | 2002 | Capacitive | 3 V | 10 MHz | OOK | - |
| 0 | 2003 | Capacitive | 1 V | 10.7 MHz | FSK | 9.6 kbps |
| 0 | 2009 | Capacitive | 1.15 V | 1 MHz | FM | 64 kbps |
| 0 | 1997 | Galvanic | 20 µA | 70 kHz | PWM | 0.9 kbps |
| 0 | 1998 | Galvanic | 3 mA | 37 kHz | FM | - |
| 0 | 2002 | Galvanic | 4 mA | 60 kHz | CPFSK | 4.8 kbps |
| 0 | 2007 | Galvanic | 1 mA | 256 kHz | BPSK | 64 kbps |

## 5.  Applications and Future Directions

IBCs can be employed to a great realm of applications, and are seen as a key component to the futuristic world of the "Internet of Things" (IoT). Example areas that can benefit from IBCs include, but are not limited to, healthcare, medical, military, emergency, and consumer electronics. The idea is that data can be transmitted within the body or even exchanged with other devices via daily actions, such as touching or handshaking. There are several scenarios for employing IBCs: (a) communication between a wearable device and an off-body device, (b) communication between two wearable devices, (c) communication between two off-body devices, and (d) communication between implanted and wearable devices.

The concept of IBCs is relatively new, and as of now no commercial IBC devices are available in the market. However, several research groups in academia and industry have been exploring and validating the applicability and performance of IBC systems. Some specific example applications for IBCs that have been already reported in the literature are given below:

•    Enabled by IBCs, a number of wearable and implantable medical sensors could transmit their sensed data to a central node placed upon the human body. Eventually, this central monitoring node can be connected by a traditional wireless link (such as Bluetooth or Wi-Fi) to a nearby PDA and/or a remote monitoring unit, such as a doctor's office or hospital 0. There, the data will be displayed, stored and post-processed as needed. A concept like this could be applied to monitor patients suffering from diabetes, asthma, heart conditions, etc.

•    In another example, two people at a conference or meeting could exchange business cards simply by shaking hands. In this case, communication can be kept private using suitable authentication and encryption technologies.

• Along the lines of secure IBCs, a person could easily unlock his house or office door just by touching the door knob.

• In 0, an IBC system was reported that transfers information about the area surrounding a human being to a cloud server.

• Recently, Intel reported a system that allows a user to share data between two devices simply by wearing an application-specific ring 0. As an example, the user can perform a copy/paste action using a particular touch interface. Placing two fingers on a laptop's touch interface will allow storing of the data to the user's ring. The user can then go to a similarly enabled device and perform the same action (i.e., place two fingers on the touch interface) in order to paste the stored data. At a desk within Intel Labs, one of the prototypes was demonstrated: a very small graphics file, an emoticon, was copied from one laptop, which was modified to include a touch-sensor, to a worn ring. It was then transferred to another similarly modified laptop via a user touching the sensor. The drawback, at least at this stage, is that the data has to be small in size – a few bytes – in order to be stored on the battery-less ring.

• Another application reported by Intel has been on picking up geographical coordinates from a mapping application, and touching a GPS device to swiftly transmit the location. According to 0, the team said they had to face challenges related to finding ways of dealing with power loss within the human body. To that end, they fine-tuned the circuits to build a safe and stable connection for data transfer.

• Touch-to-print is another reported application. This implies the ability to store the identity (ID) of a document for printing within an accessory (e.g., a bracelet). The user can then go to an IBC-enabled printer and touch the bracelet so the document would print out there.

• In another case, IBC devices could be embedded into medicine bottles and include details on the medicine's characteristics. If the user touches the wrong medicine, an alarm would be triggered. Using IBCs, false alarms that could occur by embedding RFID tags into medicine bottles and having a user in close proximity, are avoided.

• IBC devices could be embedded into clothes and other consumer items. By touching items and products they are interested in, consumers/shoppers could get more in-depth information (such as pictures, pricing, availability, etc.) directly on their PDA device.

• Another example is an IBC-enabled PC and mouse. The PC could be configured to the user's specifications simply by having the user touch the mouse.

• IBC devices integrated into human-operated tools equipped could verify the user and confirm his suitability to utilize and operate the tool.

Miniaturization of IBC devices, fabrication using flexible and unobtrusive conductive materials, and secure transmission protocols are key to transitioning IBCs into real-world applications. Making wearable technology more natural to use, unobtrusive and fashionable is another critical aspect. Concurrently, there is still a great need to investigate the properties of the human body and model biological tissues as a transmission medium for IBCs. In doing so, it will be much easier to optimize the coupling method, frequency, current/voltage amplitude, electrode size and positioning etc.

Overall, IBC has exciting prospects for making wearable electronics much more practical and useful in the near future.

# References

[1] K.S. Nikita and A. Kiourti, "Mobile communication fields in biological systems," in Electromagnetic Fields in Biological Systems, J.C. Lin (Ed.), CRC Press, 2011.

[2] A. Kiourti and K.S. Nikita, "Antennas and RF communication," in Handbook of Biomedical Telemetry, K.S. Nikita (Ed.), Wiley–IEEE Press, 2014.

[3] Z. Wang, J.L. Volakis and A. Kiourti, "Embroidered antennas for communication systems," in Electronic Textiles: Smart Fabrics and Wearable Technology, Woodhead Publishing, 2015.

[4] https://www.apple.com/watch/

[5] http://www.hexoskin.com

[6] http://press.ralphlauren.com/polotech/

[7] http://www.vitaljacket.com/

[8] http://www.sensoriafitness.com

[9] https://jawbone.com/

[10] S. Salman, Z. Wang, E. Colebeck, A. Kiourti, E. Topsakal, and J.L. Volakis, "Pulmonary edema monitoring sensor with integrated Body-Area Network for remote medical sensing," IEEE Transactions on Antennas and Propagation, vol. 62, issue 5, pp. 2787–2794, May 2014.

[11] A. Kiourti, J.R. Costa, C.A. Fernandes, and K.S. Nikita, "A broadband implantable and a dual-band on-body repeater antenna: design and transmission performance," IEEE Transactions on Antennas and Propagation, vol. 62, issue 6, pp. 2899–2908, Jun. 2014.

[12] J. Bito, J.G. Hester, and M.M. Tentzeris, "Ambient RF energy harvesting from a two-way talk radio for flexible wearable wireless sensor devices utilizing inkjet printing technologies," IEEE Transactions on Microwave Theory and Techniques, vol. 63, issue 12, pp. 4533-4543, Dec. 2015.

[13] R. Abbasi-Kesbi and A. Nikfarjam, "A mini wearable wireless sensor for rehabilitation applications," in Proc. 3$^{rd}$ International Conference on Robotics and Mechatronics, pp. 618-622, Oct. 2015.

[14] M. Seyedi, B. Kibret, D.T.H. Lai, and M. Faulkner, "A survey on intra-body communications for body area network applications," IEEE Transactions on Biomedical Engineering, vol. 60, issue 8, pp. 2067-2079, Aug. 2013.

[15] T.G. Zimmerman, "Personal Area Networks: Near-field intrabody communication," IBM. Syst. K., vol. 35, no. 3, pp. 609-617, 1996.

[16] M.A. Estudillo, D. Naranjo, L.M. Road, and L.J. Reina, "Intrabody communications as an alternative proposal for biomedical wearable systems. In Cavado, Ave (eds): Handbook of Research on Developments in e-Health and Telemedicine: Technological and Social Perspectives. Portugal: IGI Global.

[17] L.M. Roa, J. Reina-Tosina, A. Callejon-Leblic, D. Naranjo, and M.A. Estudillo-Valderrama, "Intrabody Communication," in Handbook of Biomedical Telemetry, K.S. Nikita (Ed.), Wiley–IEEE Press, 2014.

[18] INCIRP, "Guidelines for Limiting Exposure to Time-Varying Electric, Magnetic, and Electromagnetic Fields (up to 300 GHz)," Health Physics, vol. 74, no. 4, pp. 494–522, 1998.

[19] Z. Lucev, I. Krois, and M. Cifrek, "Intrabody Communication in Biotelemetry," in Wearable and Autonomous Systems, A. Lay-Ekuakille (Ed.), Springer-Verlag Berlin Heidelberg, 2010.

[20] A. Kiourti, C. Lee, and J.L. Volakis, "Fabrication of Textile Antennas and Circuits with 0.1mm Precision," IEEE Antennas and Wireless Propagation Letters, vol. 25, pp. 151-153, 2016.

[21] A. Kiourti and J.L. Volakis, "High–Geometrical–Accuracy Embroidery Process for Textile Antennas with Fine Details," IEEE Antennas and Wireless Propagation Letters, vol. 14, pp. 1474-1477, 2015.

[22] Y. Kim, B. Lee, S. Yang, I. Byun, I. Jeong, and S.M. Cho, "Use of copper ink for fabricating conductive electrodes and RFID antenna tags

by screen printing," Current Applied Physics, vol. 12, issue 2, pp. 473-478, Mar. 2012.

[23] K. Ito, M. Takahashi, and K. Fujii, "Transmission mechanism of wearable devices using the human body as a transmission channel," in Antennas and Propagation for Body-Centric Wireless Communications, P. Hall and Y. Hao (Eds.), Artich House, 2012.

[24] D. Liu and C. Svensson, "Power consumption estimation in CMOS VLSI chips," IEEE Journal on Solid-State Circuits, vol. 29, no. 6, pp. 663-670, 1994.

[25] C. Chakraborty, B. Gupta, and S.K. Ghosh, "A review on telemedicine-based WBAN framework for patient monitoring, Journal of Telemedicine and E-Health, vol. 19, no. 8, pp. 619-626, 2013.

[26] M.A. Callejon, D. Naranjo-Hernandez, J. Reina-Tosina, and L.M. Roa, "A comprehensive study into intrabody communication measurements," IEEE Transactions on Instrumentation and Measurements, vol. 62, no. 9, pp .2446-2455, 2013.

[27] Z. Nie, J. Ma, Z. Li, H. Chen, and L. Wang, "Dynamic propagation channel characterization and modeling for human body communication," Sensors, vol. 12, no. 12, pp. 17569-17587, 2012.

[28] Z.L. Vasic, I. Krois, and M. Cifrek, "Effect of body positions and movements in a capacitive intrabody communication channel from 100 kHz to 100 MHz, " in Proc. IEEE International Instrumentation and Measurement Technology Conference, pp. 2791–2795, 2012.

[29] M. Seyedi and D.T.H. Lai, "Effect of limb joints and limb movement on intrabody communications for body area network applications," Journal of Medicine and Biology Engineering, vol. 34, no. 3, pp. 276-283, 2014.

[30] M.S. Wegmüller, "Intra-Body Communication for Biomedical Sensor Networks," PhD thesis, Diss. ETH No. 17323, ETH Zürich, Switzerland, 2007.

[31] T.G. Zimmerman, "Personal Area Networks (PAN): Near-field intrabody communication," MIT Media Laboratory, Cambridge, 1995.

[32] M. Fukumoto and Y. Tonomura, "Body coupled FingeRing: Wireless wearable keyboard," in Proc. Conference on Human Factors in Computing Systems, pp. 147–154, 1997.

[33] K. Partridge, B. Dhlquist, A. Veiseh, A. Cain, A. Foreman, J. Goldberg, and G. Borriello, "Empirical measurements of intrabody communication performance under varied physical configurations," in Proc. 14th annual ACM symposium on User interface software and technology, Orlando, Florida, USA, pp. 183–190, 2001.

[34] K. Fujii, K. Ito, and S. Tajima, "Signal propagation of wearable computer using human body as transmission channel," in Proc. International Symposium on Antennas and Propagation, pp. 512–515, 2002.

[35] K. Hachisuka, A. Nakata, T. Takeda, K. Shiba, K. Sasaki, H. Hosaka, and K. Itao, "Development of wearable intra-body communication devices," Sensors and actuators A: Physical, vol. 105, pp. 109–115, 2003.

[36] Z. Lucev, I. Krois, and M. Cifrek, "A multichannel wireless EMG measurement system based on intrabody communication," in Proc. XIX IMEKO World Congress on Fundamental and Applied Metrology, Lisbon, Portugal, pp. 1711–1715, 2009.

[37] T. Handa, S. Shoji, S. Ike, S. Takeda, and T. Sekiguchi, "A very low-power consumption wireless ECG monitoring system using body as a signal transmission medium," in Proc. International Conference on Solid-State Sensors and Actuators, Chicago, USA, pp. 1003–1006, 1997.

[38] D.P. Lindsey, E.L. McKee, M.L. Hull, and S.M. Howell, "A new technique for transmission of signals from implantable transducers," IEEE Transactions on Biomedical Engineering, vol. 45, pp. 614–619, 1998.

[39] M. Oberle, "Low power system-on-chip for biomedical application," PhD thesis Diss. ETH No. 14509, ETH Zürich, Suitzerland, 2002.

[40] Y. Song, Q. Hao, and K. Zhang, "The simulation method of the galvanic coupling intra-body communication with different signal transmission paths," IEEE Transactions on Instrumentation and Measurements, vol. 60, pp. 1257–1266, 2011.

[41] N. Cho, J. Yoo, S.J. Song, J. Lee, S. Jeon, and H.J. Yoo, "The human body characteristics as a signal transmission medium for intrabody communication," IEEE Transactions on Microwave Theory and Techniques, vol. 55, pp. 1080–1085, 2007.

[42] M.S. Wegmueller, A. Kuhn, J. Froehlich, M. Oberle, N. Felber, N. Kuster, et al., "An attempt to model the human body as a communication channel," IEEE Transactions on Biomedical Engineering, vol. 54, pp. 1851–1857, 2007.

[43] Y. Song, G. Yang, Q. Hao, M. Wang, "The simulation of electrostatic coupling intra-body communication based on finite-element models," in Proc. SPIE, vol. 7157, pp. D1–D8, 2009.

[44] S.W. Franklin and S.E. Rajan, "Personal area network for biomedical monitoring systems using human body as a transmission medium," International Journal of Bio-Science and Bio-Technology, vol. 2, no. 2, Jun. 2010.

[45] T. Nakagawa, M. Utsunomiya, S. matsumoto, S. nonomura, T. Iino, T. Minotani, T. Ishihara, K. Ochiai, M. Shinagawa and Y. Kado, "Touch and Step Navigation: RedTacton application," in Proc. 8th Int. Conf. Ubiquitous Computing, 2006.

[46] http://www.intelfreepress.com/news/investigating-human-body-communications-wearables/8576/

[47] J. Bito, J.G. Hester, and M.M. Tentzeris, "Ambient RF energy harvesting from a two-way talk radio for flexible wearable wireless sensor devices utilizing inkjet printing technologies," IEEE Transactions on Microwave Theory and Techniques, vol. 63, issue 12, pp. 4533-4543, Dec. 2015.

[48] K. Partridge, B. Dhlquist, A. Veiseh, A. Cain, A. Foreman, J. Goldberg, and G. Borriello, "Empirical measurements of intrabody communication performance under varied physical configurations," in

Proc. 14th annual ACM symposium on User interface software and
technology, Orlando, Florida, USA, pp. 183–190, 2001.

# CHAPTER EIGHT

## Flexible Pentagonal Monopole Antenna

Hussein Q. AL-Fayyadh, Ali I. Hammoodi, Haider M. AlSabbagh, Hussain Al-Rizzo

*Department of Systems Engineering, University of Arkansas at Little Rock, Little Rock, AR, USA*

## Abstract

A flexible pentagonal monopole antenna is proposed to operate at multiple frequencies (1.57, 3.5, 5.2, 5.5, 5.8) GHz. The antenna is suitable for GPS, WLAN, WiMaX and ISM band applications. The antenna is printed on Kapton polyimide substrate with (49 × 34) mm dimensions and fed by a Coplanar Waveguide Transmission (CPW) line. Results show that the proposed antenna is capable of radiating efficiently at the claimed frequencies with good impedance matching.

**Keywords:** Coplanar waveguide, flexible antenna, Kapton polyimide, monopole antenna.

## 1. Introduction

Recently, flexible electronic devices have attracted a large number of research activities due to their promised applications in different fields such as military, medical, scientific, firefighting, personal communication and Radio Frequency Identification (RFID) [1]. These electronic devices need to be integrated with flexible antennas to provide wireless communication at certain allocated frequency bands. The flexible antenna must possess specific performance characteristics, such as lightweight, low-profile, compactness as well as mechanically robustness. These properties must be taken into consideration when designing a flexible antenna. In this chapter, the printed monopole type is chosen due to its interesting performance such as omni-directional radiation pattern, large impedance bandwidth, and high efficiency [2]. On the other hand, there are different types of substrates that can be used for flexible applications such as textile, paper, fluidic, polymer and synthesized. The Kapton polyimide has been chosen in this research since it exhibits a good balance between physical, chemical, and electrical properties with a low loss tangent over a wide frequency range. Also, Kapton offers a very low profile (25 μm) yet, very robust substrates with a tensile strength of 165MPa at 73°F, a dielectric strength of 3500–7000 volts/mil, and a temperature rating of 65 to 150°C [3]. The simulation of the proposed structure is done using the High Frequency Structure Simulator (HFSS) V.15, which is based on the Finite Element Method (FEM) [4].

## 2. Antenna Design

The pentagonal-shaped monopole has some attractive features which make it suitable for many wireless communication systems. A Kapton polyimide substrate is chosen to provide flexibility and mechanically robustness simultaneously. It has a dielectric constant of 3.4 and a loss tangent of 0.002 with thickness 25μm producing an ultra-low profile antenna. Our design has been inspired by the antenna structures reported in [5]. The monopole antenna is fed by a CPW transmission line to offer a single-layer fabrication process. Both the radiating

element and the CPW are printed on the same side of a (49 mm ×
34 mm) Kapton polyimide substrate. The antenna geometry and
dimensions are depicted in Fig. 1.

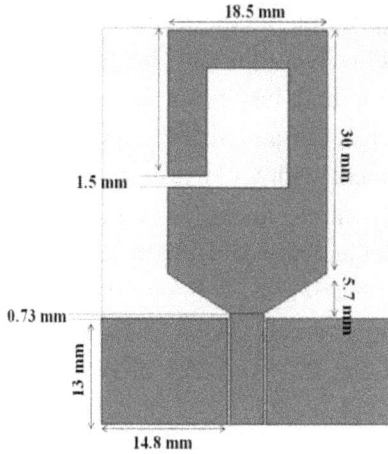

Figure 1: Proposed antenna geometry and dimensions.

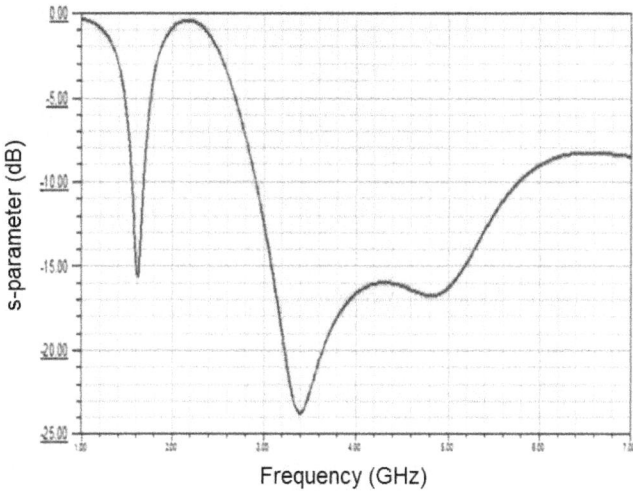

Figure 2: Simulated reflection coefficients $S_{11}$ of the proposed antenna.

## 3. Antenna Fabrication

Flexible conductive materials are not widely utilized, but some methods have been used to make flexible conductors for antennas. Inkjet printing methods have been industrialized to allow conductive, silver ink to be deposited on different substrates. This method has been used to make antennas on paper substrates, Kapton and liquid crystal polymer (LCP) [6]. The radiating element with CPW are printed on the same side of 25µm Kapton polyimide substrate using a material inkjet printer. Afterward, it is subjected to a 10 hour of thermal annealing via an industrial oven heating to evaporate the ink solvent. Then, the SMA (SubMiniature version A) terminals were attached to the radiating element and the two sides of ground plane as depicted in Fig. 3.

Figure 3: Proposed antenna prototype.

## 4. Results and Discussion

The reflection coefficient $S_{11}$ of the fabricated antenna is measured using a microwave network analyzer. The measured results are compared with those obtained from the simulator in Fig. 2. As can be seen from Fig. 4, a good agreement has been achieved between simulation and measurements.

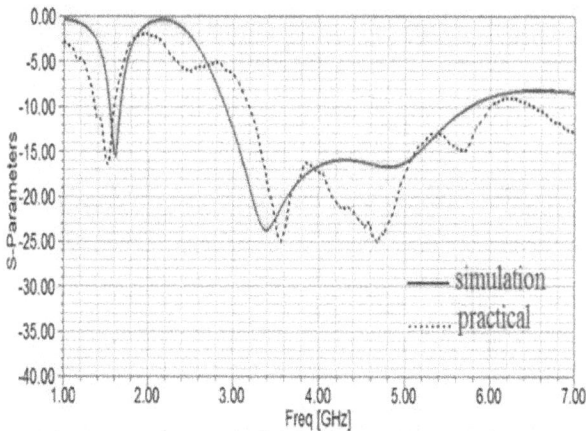

Figure 4: $S_{11}$ for the simulation /prototype antenna.

The far-field radiation patterns of the E and H planes were measured inside the University of Arkansas at Little Rock's anechoic chamber. The Antenna Under Test (AUT) was placed on an ETS Lindgren 2090 positioner and aligned to a horn antenna with a modifiable polarization where E and H planes cuts can be obtained. Radiation patterns of the fabricated antenna at (1.57, 3.5, 5.2, 5.8) GHz on the YZ plane (E-plane) and XZ plane (H-plane) are depicted in Fig. 5. It can be noted that the monopole has an omni-directional radiation pattern at GPS and WiMaX bands, and the antenna radiation pattern is close to omni-direction at WLAN and ISM bands.

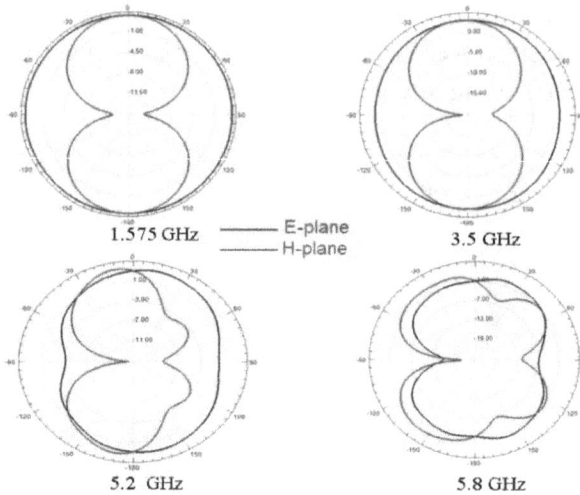

Figure 5: YZ (E-plane) and XZ (H-plane) radiation patterns for the proposed antenna at different frequencies.

## 5. Flexibility Tests

Since the proposed antenna may experience bending while conforming on specific objects during operation, flexibility tests are conducted to ensure performance reliability. Resonant frequencies and return loss need to be calculated since they are prone to shift/decrease due to impedance mismatch and change in the actual electrical length of the radiating element [1].

First, simulation of bending is performed using HFSS. Then, the measurements are done by bending the proposed antenna over a cylindrical foam, and finally, the reflection coefficients are measured as depicted in Fig. 6.

(a)

(b)

Figure 6: Flexibility testing (a) simulation.(b) practical.

The results show that the values of $S_{11}$ are slightly affected due to bending at the desired operating frequencies. Moreover, the radiation characteristics also have shown some changes due to the bending as illustrated in Figs. 7 and 8 respectively.

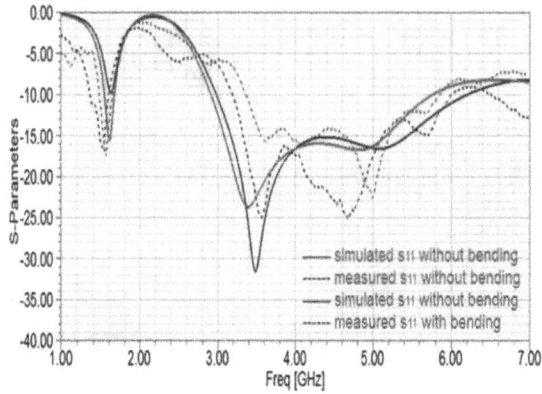

Figure 7 : Reflection cofficient (simulated / measuresd, with / without bending) of the proposed antenna for large range of frequencies.

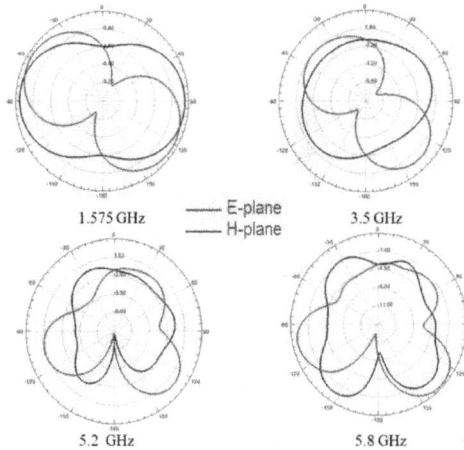

Figure 8: YZ (E-plane) and XZ (H-plane) radiation pattern for the proposed antenna under bending test at different resonant frequency.

The antenna gain is also measured with and without bending at four resonant frequencies, and the results are summarized in Table 1.

Table 1

| Frequency (GHz) | Antenna gain without bending (dB) | Antenna gain with bending, (dB) |
|---|---|---|
| 1.575 | 1.8 | 1.4 |
| 3.5 | 3.1 | 2.64 |
| 5.2 | 3.6 | 2.9 |
| 5.8 | 1.27 | 2.1 |

The results show that the boresight gain slightly decreased with bending (except at 5.8 GHz, which in fact has increased). This suggests that the proposed antenna is suitable for flexible applications.


## 6. Conclusion

This chapter presented the design of a flexible antenna structure operating at multiple frequencies covering WLAN, WiMaX, GPS, and ISM applications. A Kapton polyimide is used as a substrate with a thickness of 25μm which provides an ultra-low profile structure. Inkjet printing is used to prototype the antenna, and there is a good agreement between the simulated and measured $S_{11}$. There is no significant performance deterioration due to bending of the antenna. The antenna gain has been measured at four different frequencies with and without bending, and no degradation was observed.

# References

[1] H. Raad, A. Abbosh, H. Al-Rizzo, D. Rucker, "Flexible and Compact AMC Based Antenna for Telemedicine Applications," *IEEE Transactions on Antennas and Propagation,* vol.61, no.2, pp.524-531, Feb. 2013

[2] Q. Luo, J. Pereira, and H. M. Salgado, "Compact printed monopole antenna with chip inductor for WLAN," *IEEE Antennas and wireless Propagation. letters*, Vol. 10, pp. 880-883, 2011.

[3] Dupont Kapton Polyimide Specification Sheet Available: www2.dupont.com/kapton.

[4] Online: Ansys High Frequency Structure Simulator http/www.ansys.com

[5] W. C. Liu, C. C. Song, S. H. Chung, and J. L. Jaw, "Strip-loaded CPW-fed pentagonal antenna for GPS/WiMAX/WLAN applications," *Microwave and Optical Technology Letters*, vol. 51, pp. 48-52, 2009.

[6] R. Quarfoth, Y. Zhou, and D. Sievenpiper, "Flexible Patch Antennas Using Patterned Metal Sheets on Silicone," *IEEE Antennas and Wireless Propagation Letters*, Vol. 14, pp. 1354 - 1357, 2015.

# CHAPTER NINE

## Near-Field Communication Wearable Antennas

Razi Iqbal

*Al-Khawarizmi Institute of Computer Science, University of Engineering and Technology, Lahore, Pakistan*

## Abstract

This chapter discusses the design of Near Field Communication (NFC) wearable antennas. It begins with an introduction to NFC technology, its attributes, architecture, comparison with other peer technologies of its class and useful NFC applications currently available in the market. The chapter examines the design of NFC wearable antennas especially for small gadgets like wristbands, watches, and rings. Considerable focus has been given to different shapes, sizes, and materials for designing these antennas. Several design examples and simulation results have been also included to provide the reader with an understanding of how different sizes, shapes and materials for designing antennas can affect the overall performance of an antenna. A detailed comparison of vendor made and custom made antennas has also been provided in this chapter.

**Keywords:** Near Field Communication, Wearable Antenna, Short Range Wireless Technology

## 1. Introduction

NFC (Near Field Communication) is a short-range wireless technology used for enabling electronic devices to communicate with each other in a close proximity of around 10cm [1]. It is a simple two-way communication technology evolved from RFID (Radio Frequency Identification). NFC is based on open and maintained standards like ISO (International Standard Organization), ECMA (European association for standardizing information and communication systems) and ETSI (European Telecommunications Standards Institute). NFC is maintained by NFC-Forum, which was formed, by Sony, Phillips and Nokia back in 2004. NFC-Forum has more than 200 members including developers, institutions, and manufacturers. NFC is low power, around 15ma and has a low data transfer rate of around 424 kbps, which is sufficient for lightweight information transfer like secure financial transactions but too slow for media streaming like video playback, etc. As NFC is not built for streaming heavy amount of data and is mainly for lightweight data transfer, the low data rate of NFC is justified [2]. Similarly, NFC is ideal for situations where an instant connection is required; for example, e- ticketing at metro stations, financial transactions at shopping malls, student and staff attendance at academic institutions and contact sharing between mobile devices. The major benefit of using NFC is its simplicity. A user can use it simply by touching a tag, and the action is performed, without worrying about the complex process going on in the background, and this is what commercial technologies are all about; bringing simple solutions for the masses [3].

### a) Characteristics of NFC
Table 1 below shows characteristics of NFC along with the comparison with its sister technologies explaining why NFC will

be the next big thing.

Table1: NFC comparison with its sister technologies

|  | NFC | Bluetooth | IrDA | RFID |
|---|---|---|---|---|
| Standard | ISO/IEC 14443A | IEEE 802.15 | IEEE 802.11 | IEEE 802.15 |
| Data Rate | 424 Kbps | 1 Mbps | > 100 Mbps | Varies |
| Range | 4-10 cm | 5-10 m | < 1 m | 10-12 m |
| Ease of Use | Very Easy | Difficult | Difficult | Easy |
| Availability | Easy | Easy | Easy | Not easy |
| Instant Pairing | Yes | No | No | Yes |
| Encryption | Yes | Yes | Yes | Yes |
| Two way communication | Yes | Yes | Yes | No |

Table 1 shows the comparison between NFC and its sister technologies. Below is a detailed overview of each attribute mentioned in Table 1:

**Standard**

NFC is based on already available Felica and ISO/IEC 14443A, B standards. This standard is already available in the market and technologies like Felica is widely used in Japan for two-way communication between the electronic devices.

**Data Rate**

NFC data rate is 424kbps, which seems a little low as compared to other similar technologies like Bluetooth, IrDA, and RFID, but since NFC is built to transfer only a small amount of information between the electronic devices so this data rate is enough to serve the purpose.

### Range
The communication range of NFC is rather less than the other technologies mentioned in Table 1. This is because NFC is a contact and contactless technology for information transfer, hence, a very low range is required for communication. Because of its low range of communication, it has high security as compared to other technologies.

### Ease of use
NFC appears to be the simplest technology to use. Anyone can use this technology without even knowing much about it. It's a simple touch and go. No configuration or pairing is required. However in other technologies like Bluetooth and IrDA, a proper pairing of devices is required before using them.

### Availability
NFC is easily available. Several NFC components are already available in the market. Large organizations are already implementing NFC-enabled systems. NFC can also be found in latest mobile phones, just like Bluetooth and IrDA. In the case of RFID, however, the availability is not very easy which limits its use.

### Instant Pairing
NFC has this instant pairing feature built into its module. The user doesn't need to worry about configuring the devices manually to pair it with other NFC-enabled devices; it's always ready to be used. However in the case of Bluetooth and IrDA, configuration and pairing of devices are required, which is time-consuming.

### Encryption

NFC data can be encrypted using AES standards, which makes the data secure.

### Two-way Communication

One of the main advantages of NFC over other similar technologies like RFID is its two-way communication. Two NFC enabled devices could communicate with each other at the same time.

### b) NFC Working Modes

NFC majorly operates in two working modes, active and passive. Both modes have their own pros and cons and are equally used for various applications in the market. Below is the detailed explanation of both the working modes:

### Active Mode

In this mode, both NFC enabled electronic devices generate their own RF (radio frequency) fields and data is transferred. For example, if two NFC-enabled mobile devices want to transfer information using NFC, they must operate in Active mode. Fig. 1 below shows information exchange between two NFC-enabled smart phones in active mode.

Two way communication

Figure 1: Mobile devices communicating in Active mode

**Passive Mode**

In this mode, only one NFC-enabled device will generate its RF field and other device powers up from this RF field. For example, if NFC reader wants to read the data from NFC tag, it will generate its RF field and a tag will power up using that field which results in data transmission from NFC tag to the reader. Fig. 2 below shows an NFC enabled smart phone reading the information from NFC tag.

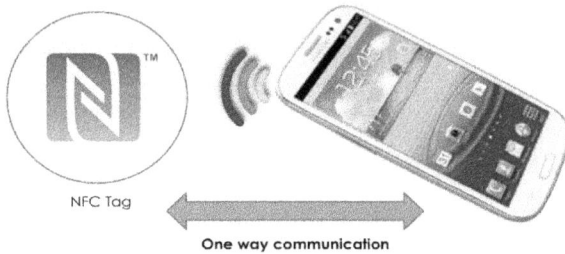

Figure 2: Mobile device communicating with NFC tag in Passive mode

## 2. NFC Wearable Antennas

The previous section has provided an overview of NFC technology by explaining its various characteristics and working modes. This section explains, in detail, NFC antennas and their capability as wearable antennas. Due to small size, NFC antennas can be an excellent choice to be used in wearable electronics; however, due to its short range, applications can be limited.

In order to understand the working of NFC wearable antennas, it is very important to have a firm grasp on the concept of mutual inductance in Near Field Communication antennas. Next section will explain in detail the concept of mutual inductance in NFC.

## 2.1 Mutual Inductance in NFC

Several studies have shown that while designing a passive antenna tag, the most prominent parameter is the area or simply the size of the coil. It not only affects the overall efficiency of the communication but also has the tendency to increase the range of the communication. There are some other parameters like shape, dimensions, number of turns of the coil, etching material, coil material and the matching circuitry that has a significant effect on overall communication between an NFC passive tag and the reader device.

NFC like its sister technologies works on the same principle of mutual inductance. Whenever electric current passes through the coil of the tag, a magnetic field is generated. An electromotive force (emf) is induced in the coil by this changing magnetic field [4].

$$\in = -N \, d\phi/dt \qquad (1)$$

In equation 1, according to Faraday's law, $\in$ is the electromotive force induced in the antenna coil, N is the number of turns of the coil and $\phi$ is the magnetic flux passing through the coil. Whenever an active antenna is brought closer to the passive antenna, a magnetic field is generated which helps in powering up the passive antenna. The power generated in the passive antenna depends on upon the area of the coil and the angle at which both active and passive coils are communicating.

The relationship between the number of turns of the coil and magnetic flux can also be defined by equation 2.

$$N \, (d\emptyset)/dt = M \, dI/dt \qquad (2)$$

In equation 2, 'M' is a constant called "Mutual Inductance Constant" and can be defined by equation 3 [5].

$$M = (N\emptyset)/I \qquad (3)$$

According to equation 3, it is evident that M depends on two factors; the number of turns of the coil, and radii of the two coils. However, detailed study of the antennas has shown that other factors like material of the coil, etching material, shape of the antenna and matching circuitry have a significant effect on the overall performance of the antenna.

## 2.2 Parameters for Designing NFC Wearable Antenna

As mentioned in the previous section, antenna efficiency and reliability depends on various parameters like size of the coil, shape of the antenna, coil material, etching material, and matching circuitry, etc. Different parameters can have a different impact on the overall communication. Let's discuss these parameters one by one using results from various experiments.

### Size / Area of the coil

One of the major factors affecting the communication in antennas is the size or the area of the coil of the antenna. Several theories are available confirming that the larger the size of the coil, the better it is to communicate. However, various studies have shown that proper proportion of several parameters is required for better communication. For example, a coil with ten turns might not work properly if the distance between each turn is not symmetric. However, a coil with six turns might give excellent results if the distance between each turn is symmetric.

In order to further understand the concept of the size of the coil and its effect on the communication, let's discuss some experimental results from various studies. Table 2 illustrates the results of experiments performed to gauge the impact of the size of the coil on antenna communication.

Table 2: Impact of coil size on the antenna communication

| Size of the Coil | Number of Turns | Number of Trails | Success Rate |
|---|---|---|---|
| 25x25 mm | 8 | 10 | 100 % |
| 25x25 mm | 11 | 10 | 100% |
| 35x10 mm | 10 | 10 | 100% |

As illustrated in Table 2, with different combinations of the size of the coil and number of turns of the coil, the success rate appears to be 100%, which is a perfect scenario for any antenna communication. However, since NFC operates at 13.56MHz frequency and if the inductance of the antenna is calculated with different size of the coil and number of turns, it should be close to 2.7µH, which depends on the NFC chip in use. For explaining the concepts here, NXP semiconductors MIFARE Ultra-light MF0ICU chip is used. The datasheet of each chip contains all the information regarding inductance of the chip. If we calculate the inductance of this chip with different coil sizes and number of turns, we can determine that different sizes can have a different impact on the inductance of the antenna, and hence, the overall communication.

Table 3: Impact of coil size on antenna communication for NXP chip

| Size of the Coil | Number of Turns | Inductance |
|---|---|---|
| 35x15 mm | 10 | 3.61 µH |
| 35x15 mm | 8 | 2.72 µH |
| 35x15 mm | 6 | 1.82 µH |

Table 3 provides an overview of how the size and the number of turns of the coil can change the inductance and hence the overall communication characteristics of the antenna. As illustrated in Table 3, best results for NXP semiconductors MIFARE Ultra-light MF0ICU chip are obtained with size 35x15 mm and eight turns since inductance appears to be 2.72 µH, which is very close to the inductance provided in the datasheet of this particular chip.

There are many other factors that affect the inductance of the antenna like permittivity, width of the coil, the number of segments, the spacing between the coil turns and thickness of the section, however, these are out of the scope of this chapter.

### Shape of the Antenna

Another important factor that can affect the overall communication of the antenna is the shape of the antenna. It is observed that antennas with different dimensions can have a different impact on the communication. Table 4 illustrates the results of various experiments performed to measure the efficiency of antennas with different shapes.

Table 4: Impact of shape of antenna on communication

| Shape of Antenna | Number of Trails | Success Rate |
|---|---|---|
| Circular | 10 | 100 % |
| Square | 10 | 100% |
| Rectangular | 10 | 100% |

Table 4 shows that a perfect antenna communication can be achieved with different coil shapes. However, to understand the concept, it is important to consider other factors affecting the antenna communication along with the shape of the coil.

As mentioned earlier, it is important to consider the inductance of the coil for antenna communication. For NFC chip, an NXP semiconductors MIFARE Ultra-light MF0ICU was used as a reference for understanding the concept presented in this chapter (inductance should be very close to 2.7μH). Several experiments were conducted to understand the effect of the coil shape along with other antenna factors. Table 5 shows results of the experiments from different studies:

Table 5: Impact of shape of antenna on communication for NXP chip

| Shape of Antenna | Number of Turns of the coil | Diameter / Size of coil | Spacing between Turns | Number of Trails | Inductance |
|---|---|---|---|---|---|
| Circular | 9 | 35 mm | 0.2 mm | 10 | 2.23 μH |
| | 10 | 35 mm | 0.2 mm | 10 | 2.75 μH |
| | 11 | 35 mm | 0.2 mm | 10 | 3.33 μH |
| Rectangular | 8 | 35 x 15 mm | 0.2 mm | 10 | 2.72 μH |
| | 9 | 35 x 15 mm | 0.2 mm | 10 | 3.17 μH |
| | 10 | 35 x 15 mm | 0.2 mm | 10 | 3.61 μH |
| | 11 | 35 x 15 mm | 0.2 mm | 10 | 2.37 μH |

As illustrated in Table 5, the shape of the coil and number of turns of the coil together have a significant effect on the coil inductance, and hence, the overall antenna communication. Experiments have shown that keeping diameter and spacing between the turns constant and changing the number of turns for a circular shaped antenna can change the inductance of the coil.

As illustrated in the table, communication quality is at its best for a circular antenna with 10 numbers of turns since the inductance is 2.75 μH, which is closest to the inductance for NXP semiconductors MIFARE Ultra-light MF0ICU. Similarly, the table also illustrates that the communication is at its best for a rectangular shaped antenna with 8 numbers of turns since the inductance is 2.72 μH, which is closest to the inductance for NXP semiconductors MIFARE Ultra-light MF0ICU.

### Other Factors Affecting the Antenna Communication

In the previous sections, discussions have shown that the two most important factors affecting the overall antenna communication based on the coil inductance are: the number of turns, and the shape of the coil. However, there are many other factors that can have significant effects on antenna communication. Below is brief overview of some of these factors:

### Coil Material

There are several materials which can be used for coiling an antenna. However, the most common are copper and aluminum. The antennas discussed in this chapter are mostly copper based antennas. The efficiency of the antenna communication can vary depending upon the use of any of these materials.

### Matching Circuitry

The input impedance of an antenna must be close to the impedance of the tag in order to minimize transmission losses. If impedance does not match, the signals will be reflected back to the passive tag and there would be no active communication. A matching circuitry might contain inductors and capacitors to match the impedance of the passive tag, which would result in efficient antenna communication.

**Winding Thickness**

Another important factor that can affect the antenna communication is the thickness of the winding. Normally, wires are winded around the loop so that each insulated winding is adjacent to each other. Here, thickness of the winding can affect the overall communication even if the winding is adjacent to each other because of various thickness levels available for winding a coil or wire.

**Spacing between the Coil Turns**

As discussed earlier, the number of turns can affect the overall antenna properties. Besides the number of turns, the spacing between the coil turns can also affect the generated magnetic flux, and hence, the overall antenna characteristics.

## 3. Conclusion

This chapter discussed the design of wearable Near Field Communication antennas. The factors that affect the overall communication of NFC wearable antennas, especially small size antennas like wristband and wristwatch antennas, were examined. The chapter also evaluated various factors that affect the antenna communication using experimental data from recent studies. Coil inductance for designing antennas for NFC is discussed in detail to provide the reader with an understanding of how inductance values play a vital role in matching the chip with the passive antenna. Various factors like coil size, antenna shape, coil material, matching circuitry, thickness of winding, and spacing between the coil turns were highlighted to provide the reader with a basic understanding of the factors that can affect the antenna communication in wearable computing.

# References

[1] Patel, Jignesh, and Badal K., "Near Field Communication - The Future Technology For An Interactive World." International Journal of Engineering Research and Science & Technology. 2(2): 55-59(2013).

[2] Marcus, Adam, Guido Davidzon, Denise Law, Namrata Verma, Rich Fletcher, Aamir Khan, and Luis Sarmenta. "Using NFC-enabled mobile phones for public health in developing countries." In Near Field Communication, 2009. NFC'09. First International Workshop on, pp. 30-35. IEEE, 2009.

[3] Fontecha, Jesus, Ramon Hervas, Jose Bravo, and Vladimir Villarreal. "An NFC approach for nursing care training." In 3rd International Workshop on Near Field Communication, pp. 38-43. IEEE, 2011.

[4] Mika, Mikko, Arto. (2009, February). "Practical implementations of passive and semi-passive nfc enabled sensors". In First International Workshop on NFC. pp. 69-74. IEEE. 2009.

[5] Jing, Wang, (2008, July). "Capacity performance of an inductively coupled near field communication system". In IEEE Antennas and Propagation Society International Symposium. pp. 1-4. IEEE. 2008.

# CHAPTER TEN

## Fiber-Based and Textile-Based Flexible Generators for Wearable Energy Harvesting and Self-Powered Sensing

Junwen Zhong, Qize Zhong, and Jun Zhou

*Wuhan National Laboratory for Optoelectronics, Huazhong University of Science and Technology, 1037 Luoyu Road, Wuhan, 430074, China*

### Abstract

Self-powered smart textiles are skyrocketing in the recent years, by virtue of the superiorities of weavability, adaptability and comfortability, compared to conventional planar and bulky structures. In this chapter, recent advances in the fabrication, structure, mechanism and electrical output performances of various fiber-based and textile-based flexible generators, including piezoelectric, triboelectric and electret generators, are summarized. Moreover, discussions will be presented regarding newfangled applications, current challenges and future directions, which can provide some useful insights.

**Keywords:** Wearable electronics, Energy harvesting, Fiber-based and Textile-based generators.

## 1. Introduction

The rapid evolution of material science and manufacturing technology brings a wide spectrum of opportunities to promote our ambient environments into an intelligent world, where wearable electronics play key roles [1-5]. Wearable electronics include kinds of electronic devices which can be "worn" on the human body and used during physical activities, [6] owning the advantages of intelligent management, convenient operation, easy portability and aesthetic appearance, *etc.* [7-9]. The movements of the human body, such as joint bending and muscle stretching, put significant strain on the wearable electronics, requiring not only flexibility, but also tolerance to stretch (10%-20% strain) for full conformability [2, 10]. Ideally, fibers and textiles are perfect building elements for wearable electronics due to their outstanding flexibility and stretchability. Compared to conventional planar and bulky structures, fiber-based and textile-based electronics (aka smart textiles) are more easily integrated with the human body, endowing them potential applications in mobile health, wireless communication and energy storage, without affecting the comfortability and normal activities [11-15].

Specifically, the power supply of the wearable electronics may limit their wide-spread usage, as the main drawback of the batteries and capacitors is the limited lifetime and inevitable environment pollution by inappropriate disposal [16, 17]. Surprisingly, the human body is a rich source of energy. It's estimated that harvesting even 1-5% mechanical energy generated by body actions, including walking, arm swinging, finger motion, and breathing, is sufficient to run many body-worn devices [18, 19]. Recently, advances in flexible generators offer general and useful routes to transfer mechanical energy into electricity, in which fiber-based and textile-based flexible generators have attracted great attention due to the advantages of excellent air permeability and maneuverability [20-22]. Herein, we summarize the recent development in various fiber-based and textile-based flexible generators, including piezoelectric, [23-27] triboelectric [28-32] and electret generators [33, 34]. The detailed fabrication, structure, mechanism and electrical output performances are introduced. Furthermore, discussions regarding newfangled applications, current challenges, and future directions are also presented, which is hoped to give some useful insights.

## 2. Energy from Human Body

The human body is a rich and direct source of energy that is inexhaustible, and will not cause any pollution. Normally, an average-sized person stores as much energy in fat as a 1000 kg battery. People use muscle to convert the stored chemical energy into positive mechanical energy, generating a sustainable power of about 100 W. This mechanical energy is used to sustain different activities, such as walking, arm swinging, finger motion, and breathing, *etc.* [19, 35]. Figure 1 indicates the total available power associated with everyday activities for a 68 kg adult, in which only about 25% is used "effectively". Because most of the human body energy is lost to wasted vibrations, it's significant to convert this power into electricity for use in wearable electronics without obviously adding the load to human body.

It is estimated that harvesting even 1-5% of mechanical energy generated by the human body will be sufficient to power a lot of body worn electronics as only the walking action can generate power of up to 67 W [36, 37].

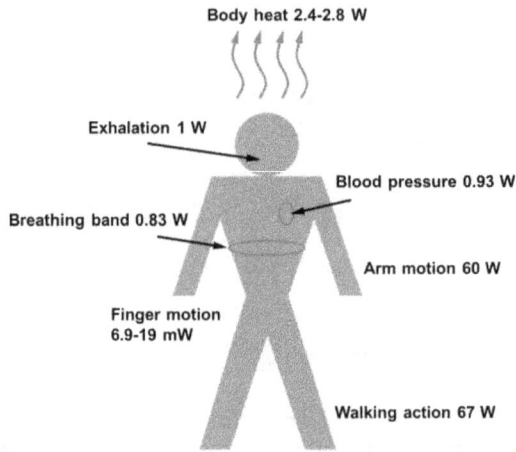

Figure 1: Total available power for everyday bodily activities [36, 37]. Copyright 2010, Royal Society of Chemistry.

Flexible generators, including piezoelectric, triboelectric and electret generators, are developed to harvest the human body mechanical energy with a wide frequency distribution (several to tens of hertz). Specifically, fiber-based and textile-based flexible generators are more easily integrated with the human body, and own the advantages of weavability, adaptability and comfortability, *etc.*

## 3. Piezoelectric Generators

### 3.1. Working Mechanism

Utilizing flexible piezoelectric generators to harvest mechanical energy attracted a great deal of attention in recent years, as these generators can directly convert mechanical energy into electricity for many integrated applications [38]. The fundamental working mechanism of piezoelectric generators is the breaking of central charges symmetry in the crystal structure under external force, which will generate a piezoelectric potential [39]. Figure 2a schematically depicts the typical wurtzite-structured zinc oxide (ZnO) crystal structure, in which the tetrahedrally coordinated Zn2+ and O2− are stacked layer-by-layer. In the original state, the charges center of cations and anions is coincident, and no piezoelectric potential is formed (Figure 2b). When the external force is applied, the structure is deformed, resulting in the separating of charges center (electric dipoles) and a piezoelectric potential between the top and bottom electrodes (Figure 2c). By connecting an external load, the piezoelectric potential will drive free electrons flow through the external circuit to achieve a new equilibrium state. Once the external force is withdrawn, the crystal structure will be recovered, and an equilibrium state is broken again, forming electric dipole and piezoelectric potential with opposite polarities (Figure 2d). In general, the variation of piezoelectric potential between the electrodes will lead to alternating currents through the external circuit.

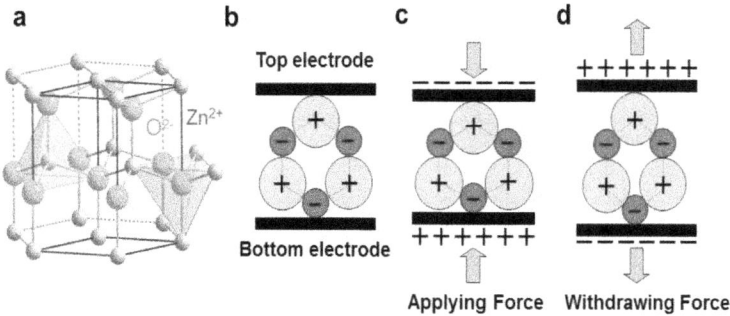

Figure 2: Working mechanism of piezoelectric generators [38, 39]. (a) Crystal structure of the wurtzite-structured ZnO. Charges center and piezoelectric potential distribution in the (b) original, (c) applying force and (d) withdrawing force states. Copyright 2009, AIP Publishing.

## 3.2. ZnO-Based Generators

As inorganic piezoelectric materials, such as ZnO [23, 24] and lead zirconate titanate (PZT) [40], are not flexible essentially, flexible piezoelectric generators are fabricated based on the coupling of piezoelectric materials and flexible substrates. Specifically, ZnO nanostructures can be grown on the fiber-shaped substrate to construct the fiber-based flexible piezoelectric generators.

For example, Qin *et al* fabricated radial ZnO nanowires on Kevlar 129 fiber surface using a hydrothermal approach. Then, tetraethoxysilane (TEOS) layers were infiltrated the ZnO nanowires to improve the mechanical performance of the fiber and the binding of nanowires, forming a composite fiber. A 300-nm-thick gold layer was coated with above composite fiber to form another composite fiber. Two fibers were entangled to form a double-fiber model system as the core for power generation, as indicated in Figure 3a. Stretching and releasing the device makes the two fibers brush with each other, and some of the gold-coated nanowires penetrated slightly into the spaces between the uncoated nanowires, resulting in the Schottky contact, as indicated Figure 3b. Piezoelectric potential is thus generated across the uncoated nanowire owing to its piezoelectric property, with the stretched surface positive ($V+$) and the compressed surface negative ($V-$), as indicated in Figure 3c. Generally, a cycled relative sliding motion between the two fibers produces power output owing to the deflection and bending of the nanowires, with about 5 pA peak short-circuit current (*Isc*) detected at each stretching-releasing cycle (blue curve in Figure 3d). When the current meter was reverse-connected, negative currents with the same amplitude were recorded (pink curve in Figure 3d), indicating that the measured output signals are generated from the generator rather than from the measurement system. This prototype introduces the concept of building a flexible, adaptable, and wearable power source which is a foundation for self-powered smart textiles [23].

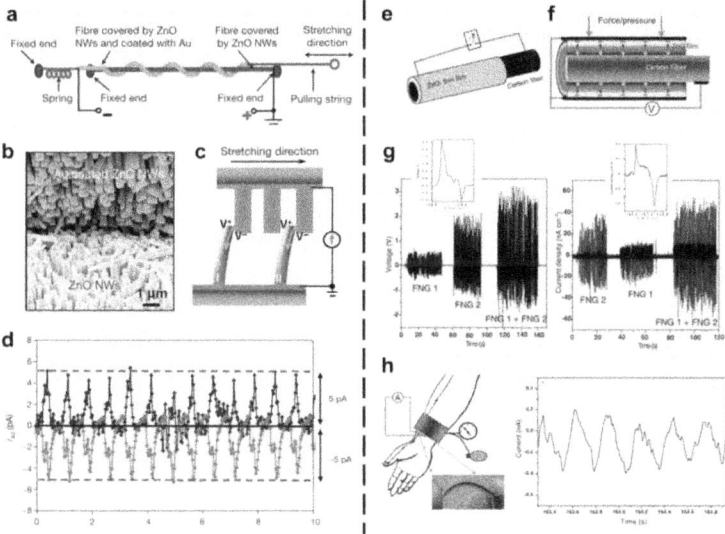

Figure 3: Piezoelectric generators based on ZnO. (a) Schematic illustration of the structure of the fiber-based generator based on ZnO nanowires. (b) SEM image at the 'teeth-to-teeth' interface of two fibers covered by nanowires, with the top one coated with Au. (c) Schematic illustration of the 'teeth-to teeth' contact between the two fibers. (d) The *Isc* of the generator with forward connecting and reverse-connecting to the measurement system [23]. Copyright 2008, Nature Publishing Group. (e) Schematic and (f) working mechanism of the fiber-based generator based on a carbon fiber coated by a ZnO thin film. (g) Two generators for a 'linear superposition' test of *Voc* and *Isc*. The *Voc* was the sum of generator 1 and generator 2 when they were connected in series; the *Isc* was the sum of the two generators when they were connected in parallel. (h) The generator acted as a self-powered sensor for monitoring the heart pulse [24]. Copyright 2011, Wiley-VCH.

Moreover, Li *et al.* developed a new approach for fabricating fiber-based flexible piezoelectric generator based on carbon fibers that are covered cylindrically by textured ZnO thin films (Figure 3e). The force or pressure applied on the textured thin films will result in a macroscopic piezoelectric potential across the thickness of the film, generating output signals (Figure 3f). The open-circuit voltage (*Voc*) was the sum of the two generators when they were connected in series (left figure in Figure 3g) and the *Isc* was also the sum of the two generators when they were connected in parallel (right figure in Figure 3g), verifying that the measured signals were indeed generated by the generators. When the generator was driven by the air pressure, maximum *Voc* of 3.2 V was reached, demonstrating the applications in non-contact conditions, such as gas pipes and oil pipes, *etc.* Moreover, this generator could act as a self-powered and ultrasensitive sensor for monitoring the heart pulse (Figure 3f), which may possibly be applied to medical diagnostics [24].

### 3.3 PVDF Based Generators

Compared to inorganic piezoelectric materials, polymer-based piezoelectric materials like polyvinylidene fluoride (PVDF) and its copolymer poly(vinylidenefluoride-co-tri-fluoroethylene) (P(VDF-TrFE)) are more suitable for fabricating fiber-based or textile-based flexible piezoelectric generators, due to their inborn advantages of flexibility, ease of processing and adequate mechanical strength, *etc.* [25-27].

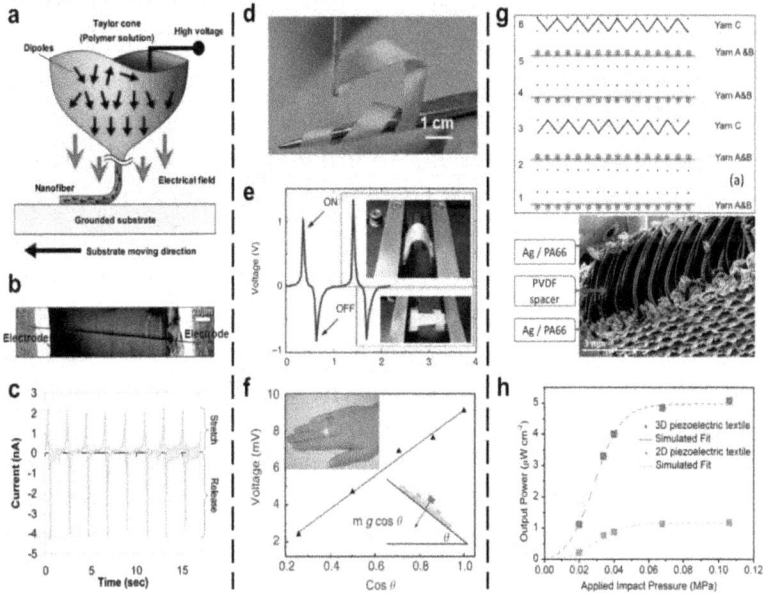

Figure 4: Piezoelectric generators based on PVDF. (a) Schematic diagram indicating the near-field electrospinning. (b) SEM image of a fiber-based generator comprised of a single PVDF nanofiber and two electrodes. (c) *Isc* for the generator [25]. Copyright 2010, American Chemical Society. (d) Photograph of a free-standing textile composed of highly aligned piezoelectric P(VDF-TrFE) fibers. (e) *Voc* for the textile-based generator under cycling bending at 1Hz. (f) Characterization of an orientation sensor based on the textile-based generator [26]. Copyright 2013, Nature Publishing Group. (g) Schematic diagram and cross-sectional SEM image of the '3D spacer' all fiber piezoelectric textile. (h) Variation of total output power as a function of applied impact pressure for 2D and 3D piezoelectric textile [27]. Copyright 2014, Royal Society of Chemistry.

Chang *et al.* used near-field electrospinning to direct-write PVDF nanofibers with high β-phase structure, as indicated in Figure 4a. The as-spun PVDF nanofibers had diameters ranging from 500 nm to 6.5 μm with variable lengths defined by the separation distance between two metallic electrodes (Figure 4b). This fiber-based generator could provide maximum *Isc* of 0.5-3 nA (Figure 4c) and maximum *Voc* of 5-30 mV, with energy conversion efficiency (order of magnitude higher than generators made of PVDF thin films). By virtue of the small size and easy fabrication, this generator can be the basis for an integrated power source in nanodevices or new self-powered textile by direct-writing nanofibers onto a large area [25].

To take a step further, Persano *et al.* developed a large area, flexible and free-standing piezoelectric textile composed of highly aligned electrospun fibers of the P(VDF-TrFE), as indicated in Figure 4d. This textile-based generator exhibited superior flexibility and mechanical robustness, and its maximum *Voc* could achieve 1.5 V under bending (Figure 4e). The wearable applications of this generator for measuring vibration/acceleration and orientation were demonstrated (Figure 4f), indicating the application opportunities in human motion monitoring and robotics [26].

A new structural design is proposed and practiced to produce all-fiber piezoelectric textile for harvesting mechanical energy. Soin *et al.* demonstrated a '3D spacer' technology based on all-fiber piezoelectric textile which consists of high β-phase PVDF monofilaments as the spacer yarn interconnected between silver-coated polyamide multifilament yarn layers acting as the top and bottom electrodes (Figure 4g). Compared to traditional 2D piezoelectric textile, this new type of 3D textile-based generator provided nearly five times larger output power density, with a maximum power density of 5.07 μW/cm2 (Figure 4h). Also, the textile can be cut into any shape and size without compromising its flexibility, and offers a simple route to integration. The high energy efficiency and mechanical durability of all-fiber based generator are highly attractive for a variety of potential applications, such as wearable electronic systems and energy harvesters driven by the ambient environment [27].

## 4. Triboelectric Generators

### 4.1. Invention and Working Mechanism

In 2012, Wang *et al* were the first to report a low-cost, simple-structured and all polymer-based flexible triboelectric generator that can convert mechanical energy into electricity effectively [41]. The basic working mechanism of triboelectric generators is coupled with triboelectrification and electrostatic induction. Triboelectrification universally exists between common materials, including conductors, semiconductors and insulators, and is conventionally known since the ancient times [38]. The mainstream opinions believe that triboelectrification is caused by the charges transfer between two materials when they contact or slide with each other. Normally, electrons will transfer from the material with low work function to the material with high work function. Empirically, polymer materials can be arranged in a table according to the number of positive charges that can be transferred, which is known as the triboelectric series [42]. Electrostatic induction is a redistribution of electrical charge in an object, caused by the influence of nearby charges. This fundamental effect is affected by the relative distance and area between the object and nearby charges [43]. Figure 5a indicates the first prototype of flexible triboelectric generator consists of polyethylene terephthalate (PET) and kapton (PI) polymer films, with metal electrodes deposited on their back sides. When the device is bent and released, triboelectrification occurs, with PET and PI capturing positive and negative charges, respectively. Simultaneously, the triboelectric charges generate equal amount but opposite signs of charges in both electrodes, according to electrostatic induction. The output currents of generators were added when connected in parallel, verifying that the measured signals are real (Figure 5b). The electrical output of the generator achieved a peak voltage of 3.3 V and current of 0.6 µA with a peak power density of about 10.4 mW/cm3 [41].

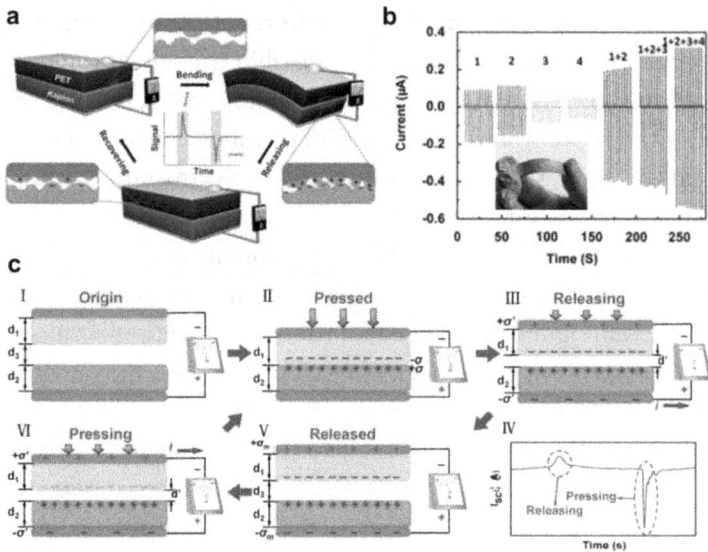

Figure 5: Working mechanism of a flexible triboelectric generator. (a) Schematic illustration of the structure and working mechanism of the first prototype of flexible triboelectric generator. (b) The total output currents of four generators connected in parallel, indicates the picture of a generator [41]. Copyright 2012, Elsevier. (c) Detailed working processes of the flexible triboelectric generator [44]. Copyright 2012, American Chemical Society.

Figure 5c illustrates the detailed working processes of the flexible triboelectric generator, which is defined as the contact-separation mode. Firstly, the generator is pressed and the two tribo-materials contact, generating triboelectric charges (I and II). As the two tribo-materials separate, the triboelectric charges in the interfacial regions are separated, which will induce charges on the surface electrodes (II and III). Thus, the electrical potential between the two electrodes is damaged, driving free electrons flow through the external circuit. In this process, electrons keep flowing until the two tribo-materials are fully released (IV and V), and the electrical potential becomes balanced. As the two tribo-materials approach again, the electrical potential undergoes the opposite variation tendency compared to the releasing process, and electrons will flow back (VI). In conclusion, periodically pressing and releasing the generator will generate alternating electrical signals [44].

## 4.2. Two Components Structured Textile-Based Flexible Triboelectric Generators

Compared to flexible piezoelectric textiles, textile-based flexible triboelectric generators possess the advantages of higher conversion efficiency, simpler fabrication process, and lower cost [28-30]. Cui *et al.* developed a cloth-based wearable triboelectric generator made of nylon and dacron fabric, and the generator had a layered structure with two components (Figure 6a). Utilizing the friction between nylon and dacron, the generator converts the mechanical energy of body motion into electrical energy naturally. With two components attaching on the experimenter's inner forearm and waist (Figure 6b), respectively, peak output current up to 150 μA was generated by the arm swinging action (Figure 6c) which could easily power an electroluminescent tube-like lamp. Moreover, this generator can be folded and kneaded, like the common clothes, which makes it possible to be integrated into clothes as a real wearable generator [28].

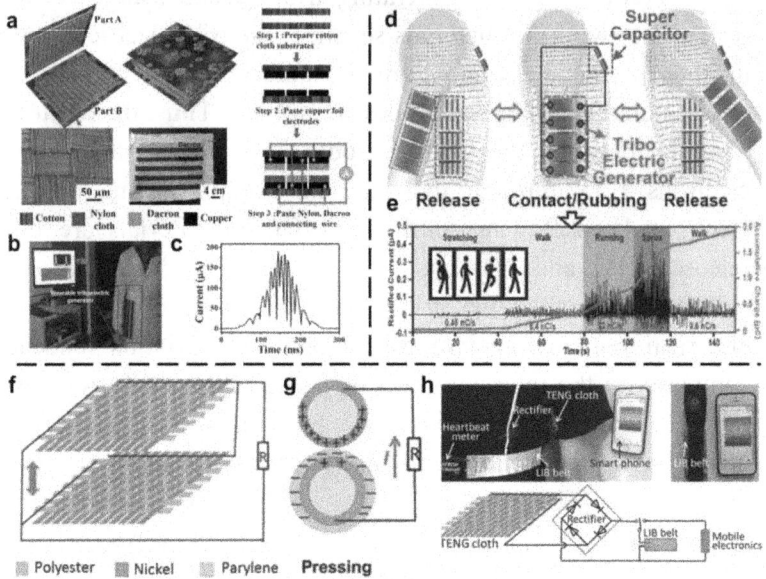

Figure 6: Two components structured textile-based flexible triboelectric generators. (a) Schematic diagram indicating the structure and fabricating processes of the cloth-based wearable triboelectric generator. (b) Photograph of a working wearable generator sewn on the clothes. (c) One current wave packet of rectified short-circuit current by the generator [28]. Copyright 2014, American Chemical Society. (d) Schematic illustration of arm swings with textile-based triboelectric generator and supercapacitor equipped. (e) Demonstration of power generation stimulated by daily body actions [29]. Copyright 2014, Wiley-VCH. Schematic illustration of the (f) structure and (g) working mechanism for the woven-structured triboelectric generator. (h) Self-charging power unit composed of textile-based triboelectric generator and a lithium-ion battery for powering a heartbeat meter with remote communication with a smart phone [30]. Copyright 2015, Wiley-VCH.

In order to store the irregular electric energy generated by the power textiles, there is a pressing need to develop integrated power generation and storage devices. Jung *et al.* reported a textile-based integrated energy device composed of a two components structured triboelectric generator and supercapacitor, which were both fabricated on a conductive carbon fabric (Figure 6d). The triboelectric generator can utilize both the vertical and horizontal friction generated between the arm and the torso, allowing it to generate power without the need for air gaps. Mechanical energy generated by daily body actions, such as stretching, walking, running and sprinting, was harvested by the generator (Figure 6e) to charge the supercapacitor. Then, the accumulated energy in the supercapacitor could also supply power to wearable sensors, like a pressure sensor. This integrated energy device is beneficial to increase the energy utilization efficiency, demonstrating new insights into the future development of wearable electronic systems [29].

Furthermore, Pu *et al.* developed a woven-structured triboelectric textile based on nickel-coated polyester fabric (Ni-cloth) and parylene-coated Ni-cloth fabric woven together, and these two triboelectric textiles were assembled together to form a generator (Figure 6f). When two components of the generator are contacted and separated, electric potential difference generated by triboelectric charges will generate current signals in the external circuit (Figure 6g). The same group also fabricated a textile-based lithium-ion battery to be integrated with a generator to form a self-charging power unit, which then was able to power up a heartbeat meter with remote communication with a smart phone (Figure 6h). As well as the aforementioned example, this prototype also verifies the viability of the whole self-charging power unit for future wearable smart electronics [30].

## 4.3. Single Component Structured Textile-Based Flexible Triboelectric Generators

The two components structure for traditional textile-based triboelectric generators makes them go against assembling within the human body, limiting the generators' applicability for human body energy harvesting. Consequently, it is highly desirable to develop a new type of textile-based energy harvesting devices that can be more easily integrated within human body [31, 32]. For example, Liu *et al.* reported a new type of 3D spacer triboelectric textile without a multilayer structure. Specifically, the triboelectric textile was based on commercially available fabrics with a three-dimensionally penetrated structure, which was composed of patterned poly(dimethylsiloxane) (PDMS) substrate with carbon nanotube (CNT) as bottom electrode, poly(ethylene terephthalate) (PET) fibers woven across the top surface and penetrated along the thickness and silver (Ag) fibers coated on the top of the PET fibers as the top electrode (Figure 7a). Therefore, the fabric is elastic in the thickness direction and could be reversibly compressed.

Similar to generators with two components, the contacting of PDMS and PET will generate triboelectric charges, and the reversible contacting and separating between these two triboelectric materials can drive electrons flow through the external circuit to form the alternating current signals (Figure 7b).

The peak short-circuit current and open-circuit voltage of the generator reached about 25 µA and 400 V (Figure 7c), respectively, with maximum peak output power density of 153.8 mW/m2. A generator was connected to power an array of 49 light emitting diodes which were connected in series (Figure 7d), vividly showing its excellent power generation ability [31].

For most of the textile-based triboelectric generators, the output power is generated from the contact between fibers. Thus, their output may be not sufficient and stable enough under a highly stretched condition. Kim *et al.* demonstrated a highly stretchable 2D triboelectric textile that was composed of woven fiber-based generator containing aluminum (Al) wire with zinc oxide (ZnO) nanowires in the core and a PDMS tube with Al foil in the shell (Figure 7e). When the triboelectric textile is stretched (Figure 7f), PDMS and ZnO act as the triboelectric materials, and the Al wire and Al foil are the two electrodes. Under the same stimulating mechanical force, the outputs increased with the area of the triboelectric textile increasing.

The peak output voltage and current reached 40 V and 210 μA, respectively, when the area got 14 × 14 cm2 (Figure 7g). The triboelectric textile was integrated to coats as a power-generating arm patch, whose generated current could reach 1 μA when bent at 60o. If the bending angle is increased to 135°, the current increases to 7 μA, lighting up the green LED during motion of the elbow (Figure 7h). The above-mentioned approach provides a promising candidate for high-performance and stable power textile toward self-powered wearable electronics in the near future [32].

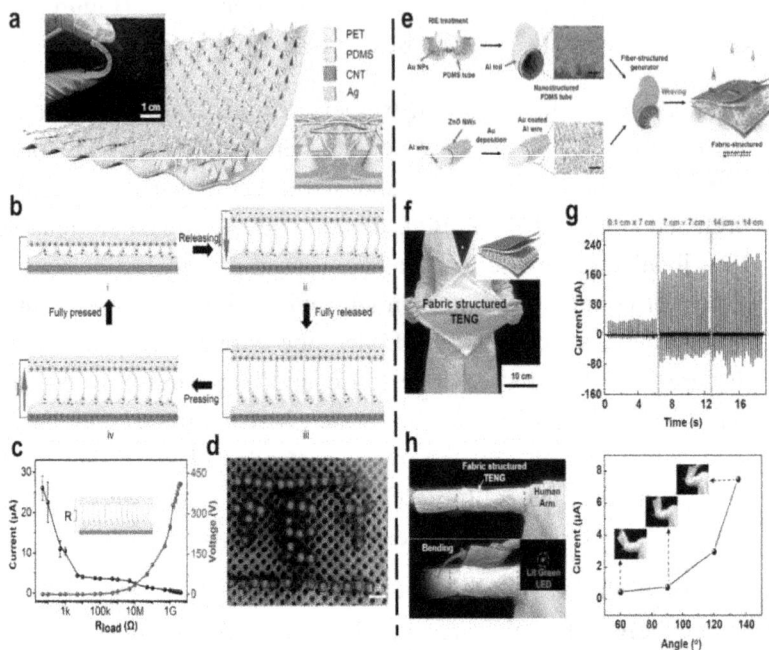

Figure 7: Single component structured textile-based flexible triboelectric generators. (a) Schematic illustration of the structure of the 3D spacer triboelectric textile. (b) Schematic illustration of the electricity generation under pressing and releasing. (c) Dependence of output current and voltage on load resistance. (d) 49 light-emitting diodes lit up by the triboelectric textile [31]. Copyright 2015, Royal Society of Chemistry. (e) Schematic diagrams of the fabrication process for the highly stretchable 2D triboelectric textile. (f) Digital picture indicating the stretchability of the triboelectric textile. (g) Output current of the triboelectric textile with the different area. (h) Output current of the 2D triboelectric textile integrated into coats as a power-generating arm patch, as a function of bending angle from 60° to 135° [32]. Copyright 2015, American Chemical Society.

## 5. Electret Generators

### 5.1 Electret and Working Mechanism of Electret Generators

An electret is a piece of dielectric material that exhibits quasi-permanent electric charges, including surface charges, space charges and dipolar charges [45], in which surface charges, and space charges are also called surplus charges (Figure 8a). For the electret film with electrode covering on one side surface, compensation charges will be generated in the electrode according to the electrostatic induction. Normally, the compensation charges (induced charges) can't surmount the barrier between the electret and electrode to be neutralized with the charges in an electret. For the time span, the entire decay time is much longer than the creating time for the charges in electret. Specifically, the lifetime can be as long as hundreds of years for the charges in excellent electret materials like polytetrafluoroethylene (Teflon, PTFE) [46], which belongs to typical fluorocarbon polymers (Figure 8b).

Usually, corona charging is adopted to inject surplus charges to the electret candidates. In this case, voltage as high as ten-odd kilovolt is applied on the corona probe. Thus, the air between the electret and probe is broken down, generating surplus charges that are mostly captured on the shallow surface of the electret (Figure 8c). The surface charge density of the electret can be confirmed by measuring the surface potential. Figure 8d shows the typical surface potential decay curves of PTFE, indicating that the maximum surface charge density ($\sigma$) for this kind electret is about 0.1-1×10-3 C/m2. Except for PTFE, fluorinated ethylene propylene (FEP), polyethylene (PE), polypropylene (PP), polyethylene terephthalate (PET) and polyimide (PI), *etc.*, are common electret organic electret materials. Besides, inorganic electret materials include silicon dioxide ($SiO2$), silicon nitride ($Si3N4$) and white mica, *etc.* [45].

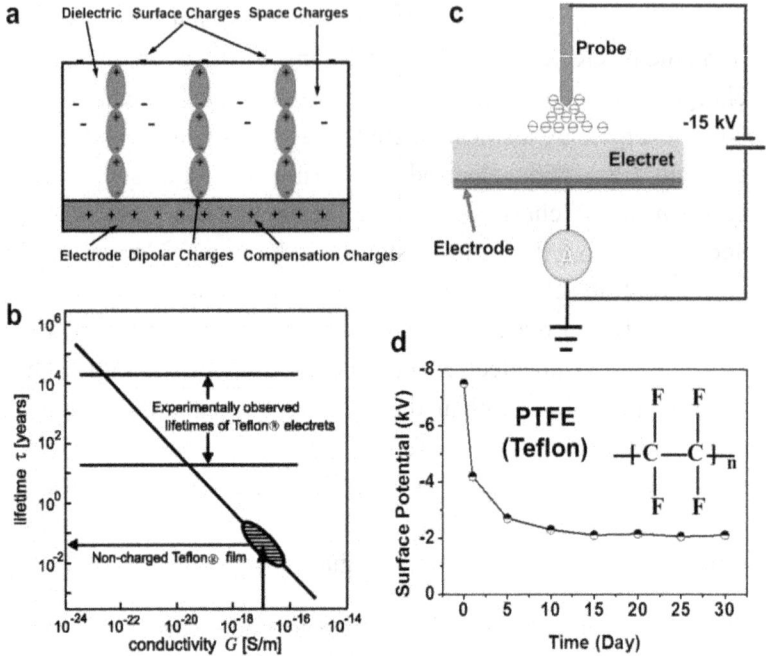

Figure 8: Electret. (a) Charges distribution in electret. (b) Electret lifetime versus electric conductivity [45, 46]. Copyright 1999, American Physical Society. (c) Schematic illustration of the corona charging. (d) Surface potential decay curves of PTFE insert shows the chemical formula of PTFE [45].

The fundamental working mechanism of electret generators is based on the electrostatic induction caused by the surplus charges of the electret materials [47, 48]. In a simplified model, the equivalent circuit of the electret generator can be regarded as a flat-panel capacitor (Figure 9a). As the inner surface of the electret is charged with surplus charges ($\sigma$), induced charges will generate in both top ($\sigma_1$) and bottom electrode ($\sigma_2$), in which $\sigma = -(\sigma_1 + \sigma_2)$. The variation of the air gap between the two components (d2) will result in the redistribution of the induced charges between the two electrodes. Thus, an electrical potential difference is generated, resulting electrons flow through the external circuit. Eventually, mechanical energy is converted into electricity. The working mechanism of the electret generator is similar to a variable-capacitance generator, except that the bias is provided by the surplus charges rather than an external voltage source [47].

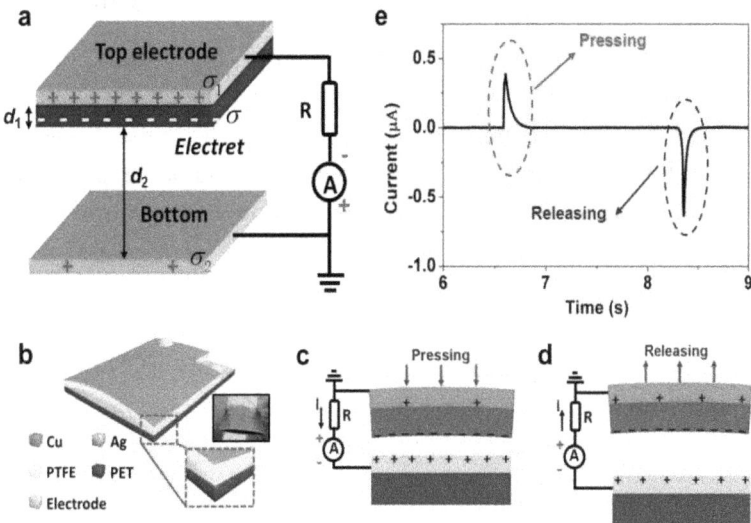

Figure 9: Working mechanism of electret generator. (a) Simplified device structure and the equivalent circuit of an electret generator. (b) Schematic diagram and digital photography of an arch-structured flexible electret generator. The equivalent circuit of this generator when it's at (c) pressing, and (d) releasing states (e). The current–time curve for this generator during a full working cycle [47]. Copyright 2013, Elsevier.

Figure 9b indicates an arch-structured flexible electret generator with PTFE as electret and copper (Cu) and Ag as the top and bottom electrode, respectively. When the generator is pressed (Figure 9c), a reduction of the air gap will lead to more induced charges in the bottom electrode, generating an instantaneous positive current (Figure 9e). Once the generator is released (Figure 9d), it will revert to its original arch shape due to its own elasticity. The air gap will increase, and more induced charges will be induced in the top electrode, resulting in an instantaneous negative current (Figure 9e) [47].

## 5.2 Fiber-Based Electret Generator for Mobile Health

As electret is injected with surplus charges beforehand, electret generator can provide stable outputs and avoid the abrasion caused by the frequent friction [33, 34]. Therefore, electret generators are appropriate for supplying sustainable power for wearable electronics. For example, Zhong *et al.* introduced a cost-effective, metal-free fiber-based electret generator that can convert biomechanical motions/vibration energy into electricity [33]. The fiber-based generator consists of two entangled modified-cotton-threads, one is carbon nanotube (CNT) coated cotton thread (CCT); the other is PTFE and carbon nanotube coated-cotton thread (PCCT) (Figure 10a).

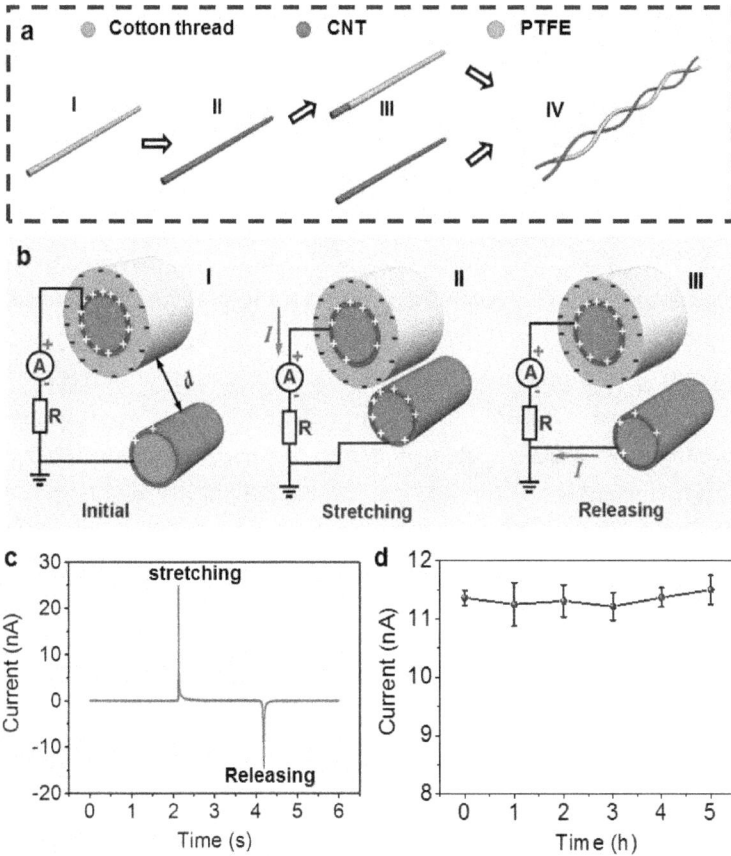

Figure 10: Fiber-based electret generator. (a) Schematic diagram illustrating the fabricating process of a fiber-based electret generator. (b) Schematic diagram illustrating the power generation mechanism of the generator, when the device is at (I) the original, (II) stretching, and (III) releasing states, respectively. (c) The corresponding output current-time curves. (d) 90000 cycles continuous power generation of the generator [33]. Copyright 2014, American Chemical Society.

Although the generator has a complex double-helix structure, it can be approximately regarded as numerous parallel-wire capacitors connected in parallel (Figure 10b-I). When the FBG is stretched (Figure 10b-II), shrinkage of the gap distance between CCT and PCCT will result in more induced positive charges in CNT electrode of CCT because of the electrostatic induction. Thus, this process produces an instantaneous positive current (Figure 10c). In the reverse case, when the generator is released (Figure 10b-III), it will revert to its original shape, and the internal gap is increased, an instantaneous negative current will be produced (Figure 10c). A generator was continuously operated for 90000 cycles, and only a small variation of peak output current is observed, indicating the highly stable power generation (Figure 10d).

Eight fiber-based electret generator were woven into fabric and connected in parallel; then fabric was sewn on a lab coat to fabricate a power textile. When the lab coat was shaken, an alternating output current would be generated (black curve in Figure 11a). The output electric signals were first rectified by a bridge rectifier (inset in Figure 11a), transforming alternating current to direct current (red curve in Figure 11a) and charging a capacitor continuously. The charged capacitor could then light up a red light LED (Figure 11b).

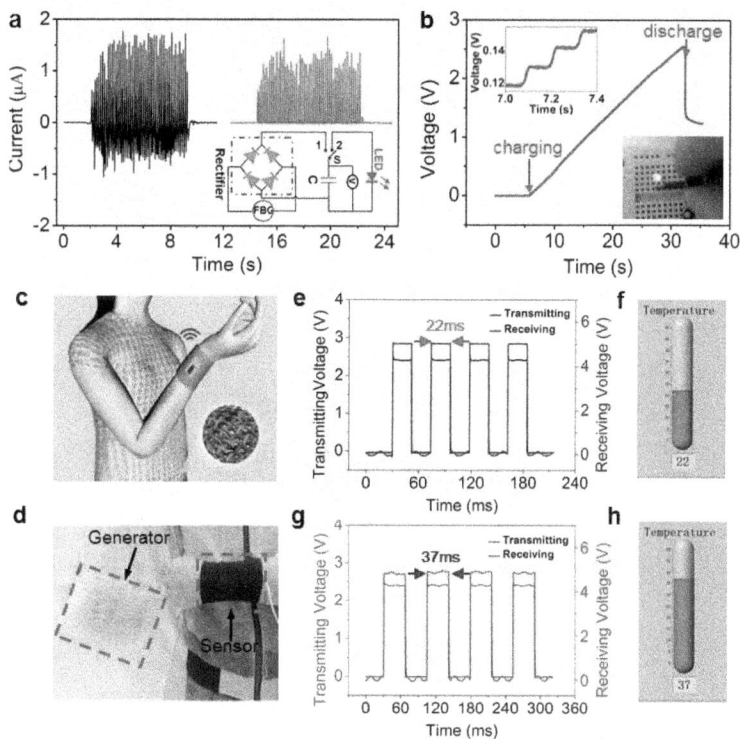

Figure 11: Power textile composed of Fiber-based electret generator. (a) The output current of the power textile without (black curve) and with (red curve) rectification when the lab coat was being shaken. (b) Voltage charging curve of a 2.2 μF commercial capacitor by the power textile. The lower inset depicts digital photography of a lighted LED powered by the charged capacitor. (c) Schematic diagram and (d) digital photography of a wireless body temperature monitor system triggered by the power textile. Modulated and demodulated signals that represent the temperatures detected by the sensor when the wristbands were positioned (e) on the desk or (g) on the human wrist. The corresponding temperature values (f) of 22 °C for room temperature and (h) 37 °C for body temperature were shown on the display screen [33]. Copyright 2014, American Chemical Society.

The power textile had successfully triggered a homemade wireless body temperature monitoring system. The working principle of the wireless body temperature monitor system is schematically shown in Figure 11c and Figure 11d. The active body temperature monitor system would detect the surrounding temperature where the wristband put on when shaking the lab coat. The modulated signals shown in Figure 11e and Figure 11g represent the detected temperatures when the wristband was positioned on the desk or on the human wrist, respectively. Meanwhile, the corresponding temperature values of 22 oC for room temperature and 37 oC for body temperature were shown on the display (Figure 11f and Figure 11h). This work establishes the first proof-of-concept that fiber-based generators can be woven into textiles and extract energy from biomechanical motions for powering mobile health systems, making the self-powered smart textile possible.

## 5.3 Stretchable Active Fiber-Based Strain Sensor

To take a step further, Zhong *et al.* developed an active fiber-based strain sensor [34], which was based on the abovementioned fiber-based electret generator. Specifically, the fiber-based electret generator was coiled around a stretchable silicone fiber to form a strain sensor with helical structure (Figure 12a). The output performance of the strain sensor depends on the generator, and the helical structure is helpful to buffer the strain.

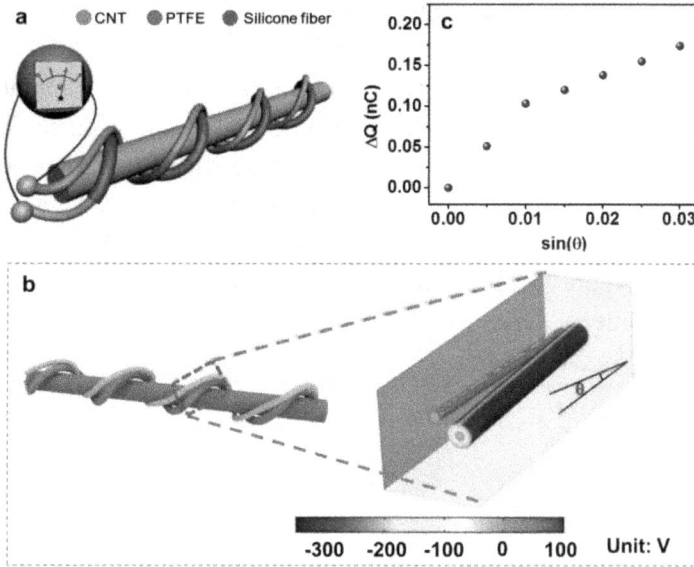

Figure 12: Active fiber-based strain sensor. (a) Schematic diagram indicating the structure of the strain sensor. (b) The finite element simulation of the working mechanism for the strain sensor. (c) The theoretical transferred charges according to the finite element simulation [34]. Copyright 2015, Wiley-VCH.

Finite element simulation is used to investigate the proposed working mechanism. Ignoring the silicon fiber which has little effect on the results, and considering the symmetry of the coiling cycle unit in the fiber-based generator, only half a unit of a single coiling cycle is constructed for simulation. Initially, two fibers have an angle of $\theta$ (Figure 12b), and the transferred charges increase with the $\sin(\theta)$ value increasing (Figure 12c). In conventional active strain sensor, like piezoelectric strain sensor, the output voltage/current of strain sensor is used to evaluate the magnitude of applied strains. However, the output current varied with stimulated frequency increasing (Figure 13a), indicating that the current/voltage data are unsuitable to evaluate this fiber-based strain sensor. The transferred charges varied linearly with the applied strains and were uninfluenced by the stimulated frequency (Figure 13b). Thus, the fiber-based strain sensor can be evaluated by measuring the integral transferred charges. The experimental results agree well with the simulation results. Besides, up to 25% strain could be detected by fiber-based strain sensor (Figure 13c), as the helix structure brings an excellent stretchability.

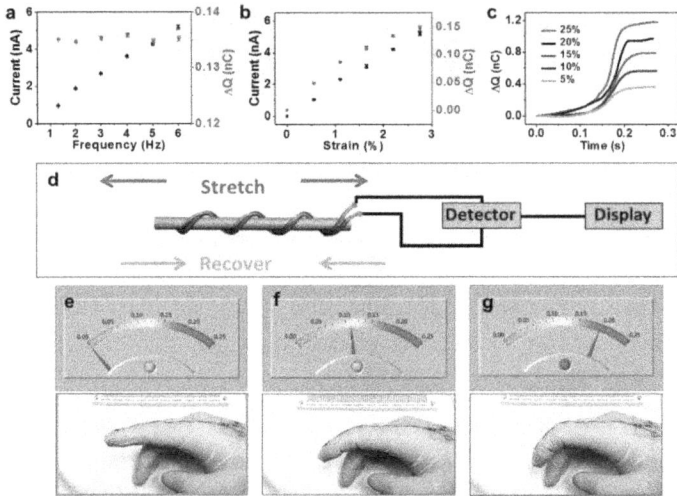

Figure 13: Active fiber-based strain sensor for detecting high degree strains and body motion. The peak value of the output currents and integral transferred charges of the strain sensor varied with (a) different stimulated frequencies for a given strain, and (b) with different stimulated strains for a given frequency. (c) The peak value of the output currents and integral transferred charges of the strain sensor varied with stimulated strains from 5% to 25%. (d) The schematic diagram for detecting finger motion states. The detecting results and digital photos show the fingers (e) did not move, (f) moved with little amplitude, and (g) large amplitude [34]. Copyright 2015, Wiley-VCH.

A fiber-based strain sensor was fixed on a finger to detect the finger motion states. Specifically, the currents generated by the finger motions were rectified, and the generating charges were stored in a capacitor, which would generate a voltage across the capacitor (Figure 13d). Then a detector was utilized to detect the voltage, and the results were shown on a screen. When the finger didn't move at all, no signal was observed (Figure 13e). If the finger moved with a little amplitude, the detector would respond (Figure 13f). Once the moving amplitude was large enough, the alert would be triggered, turning green to red on the screen (Figure 13g). This study indicated that the fiber-based strain sensor can be used as a self-powered active strain sensor for detecting human motion, and has potential application in personal health and environmental vibration detection.

## 6. Conclusion and Outlook

In conclusion, various fiber-based and textile-based piezoelectric, triboelectric and electret generators have been proposed and practiced to convert mechanical energy into electricity or act as self-powered electromechanical sensors. Generally, all these fiber-based and textile-based generators possess the advantages of flexibility, excellent assembly with the human body, weavability and stretchability, *etc.* We predict that the future work in fiber-based and textile-based generators may be in the following directions: (1) the outputs of the generators should be further improved, as they can only power up some low-power consumption electronics presently; (2) the durability of the generators must be an essential factor for ensuring practical applications, that is, the generators should be washable. (3) Power management system should be integrated within the generators and the energy storage devices, in order to more effectively use the generated power.

# References

[1] Stoppa, M.; Chiolerio, A. Wearable Electronics and Smart Textiles: A Critical Review. *Sensors* 2014, 14, 11957.

[2] Nathan, A.; Ahnood, A.; Cole, M. T.; Sungsik, L.; Suzuki, Y.; Hiralal, P.; Bonaccorso, F.; Hasan, T.; Garcia-Gancedo, L.; Dyadyusha, A.; *et al*. Amaratunga, G.; Milne, W. I. Flexible Electronics: The Next Ubiquitous Platform. *P. IEEE* 2012, 100, 1486-1517.

[3] Someya, T. Flexible electronics: Tiny Lamps to Illuminate the Body. *Nat. Mater.* 2010, 9, 879-880.

[4] Hang, Q.; Oleg, S.; Maksim, S. Flexible Fiber Batteries for Applications in Smart Textiles. *Smart Mater. Struct.* 2015, 24, 025012.

[5] Zeng, W.; Shu, L.; Li, Q.; Chen, S.; Wang, F.; Tao, X. M. Fiber-Based Wearable Electronics: A Review of Materials, Fabrication, Devices, and Applications. *Adv. Mater.* 2014, 26, 5310-5336.

[6] Lam Po Tang, S. Recent Developments in Flexible Wearable Electronics for Monitoring *Applications. T. I. Meas. Control* 2007, 29, 283-300.

[7] Li, L.; Au, W.; Ding, F.; Hua, T.; Wong, K. S. Wearable Electronic Design: Electrothermal Properties of Conductive Knitted Fabrics. *Text. Res. J.* 2014, 84, 477-487.

[8] Bauer, S. Flexible Electronics: Sophisticated Skin. *Nat. Mater.* 2013, 12, 871-872.

[9] Liu, Z.; Xu, J.; Chen, D.; Shen, G. Flexible Electronics Based on Inorganic Nanowires. *Chem. Soc. Rev.* 2015, 44, 161-192.

[10] Yu, G.; Hu, L.; Vosgueritchian, M.; Wang, H.; Xie, X.; McDonough, J. R.; Cui, X.; Cui, Y.; Bao, Z. Solution-Processed Graphene/$MnO_2$ Nanostructured Textiles for High-Performance Electrochemical Capacitors. *Nano Lett.* 2011, 11, 2905-2911.

[11] Fu, Y.; Cai, X.; Wu, H.; Lv, Z.; Hou, S.; Peng, M.; Yu, X.; Zou, D. Fiber Supercapacitors Utilizing Pen Ink for Flexible/Wearable Energy Storage. *Adv. Mater.* 2012, 24, 5713-5718.

[12] Pan, C.; Li, Z.; Guo, W.; Zhu, J.; Wang, Z. L. Fiber-Based Hybrid Nanogenerators for/as Self-Powered Systems in Biological Liquid. *Angew. Chem. Int. Edit.* 2011, 50, 11192-11196.

[13] Gao, Z.; Song, N.; Zhang, Y.; Li, X. Cotton-Textile-Enabled, Flexible Lithium-Ion Batteries with Enhanced Capacity and Extended Lifespan. *Nano Lett.* 2015. 15, 8194–8203.

[14] Li, X.; Lin, Z.-H.; Cheng, G.; Wen, X.; Liu, Y.; Niu, S.; Wang, Z. L. 3D Fiber-Based Hybrid Nanogenerator for Energy Harvesting and as a Self-Powered Pressure Sensor. *ACS Nano* 2014. 8, 10674–10681

[15] Liu, L.; Yu, Y.; Yan, C.; Li, K.; Zheng, Z. Wearable Energy-Dense and Power-Dense Supercapacitor Yarns Enabled by Scalable Graphene-Metallic Textile Composite Electrodes. *Nat. Commun.* 2015, 6. 7260.

[16] Zhong, J.; Zhu, H.; Zhong, Q.; Dai, J.; Li, W.; Jang, S. H.; Yao, Y.; Henderson, D.; Hu, Q.; Hu, L.; Zhou, J. Self-Powered Human-Interactive Transparent Nanopaper Systems. *ACS Nano* 2015, 9, 7399-7406.

[17] Lee, M.; Bae, J.; Lee, J.; Lee, C. S.; Hong, S.; Wang, Z. L. Self-Powered Environmental Sensor System Driven by Nanogenerators. *Energy Environ. Sci.* 2011, 4, 3359-3363.

[18] Donelan, J. M.; Li, Q.; Naing, V.; Hoffer, J. A.; Weber, D. J.; Kuo, A. D. Biomechanical Energy Harvesting: Generating Electricity During Walking with Minimal User Effort. *Science* 2008, 319, 807-810.

[19] Kuo, A. D. Harvesting Energy by Improving the Economy of Human Walking. *Science* 2005, 309, 1686-1687.

[20] Lee, S.; Ko, W.; Oh, Y.; Lee, J.; Baek, G.; Lee, Y.; Sohn, J.; Cha, S.; Kim, J.; Park, J.; Hong, J. Triboelectric Energy Harvester Based on Wearable Textile Platforms Employing Various Surface Morphologies. *Nano Energy* 2015, 12, 410-418.

[21] Zhou, T.; Zhang, C.; Han, C. B.; Fan, F. R.; Tang, W.; Wang, Z. L. Woven Structured Triboelectric Nanogenerator for Wearable Devices. *ACS Appl. Mater. Inter.* 2014, 6, 14695-14701

[22] Post, E. R. Electrostatic Power Harvesting in Textiles. *Proc. ESA Annual Meeting on Electrostatics* 2010, Paper G1.

[23] Qin, Y.; Wang, X.; Wang, Z. L. Microfibre-Nanowire Hybrid Structure for Energy Scavenging. *Nature* 2008, 451, 809-813.

[24] Li, Z.; Wang, Z. L. 3. *Adv. Mater.* 2011, 23, 84-89.

[25] Chang, C.; Tran, V. H.; Wang, J.; Fuh, Y.-K.; Lin, L. Direct-Write Piezoelectric Polymeric Nanogenerator with High Energy Conversion Efficiency. *Nano Lett.* 2010, 10, 726-731.

[26] Persano, L.; Dagdeviren, C.; Su, Y.; Zhang, Y.; Girardo, S.; Pisignano, D.; Huang, Y.; Rogers, J. A. High Performance Piezoelectric Devices Based on Aligned Arrays of Nanofibers of Poly(vinylidenefluoride-co-trifluoroethylene). *Nat. Commun.* 2013, 4, 1633.

[27] Soin, N.; Shah, T. H.; Anand, S. C.; Geng, J.; Pornwannachai, W.; Mandal, P.; Reid, D.; Sharma, S.; Hadimani, R. L.; Bayramol, D. V.; *et al.* Novel "3-D Spacer" All Fibre Piezoelectric Textiles for Energy Harvesting Applications. *Energy Environ. Sci.* 2014, 7, 1670-1679.

[28] Cui, N.; Liu, J.; Gu, L.; Bai, S.; Chen, X.; Qin, Y. Wearable Triboelectric Generator for Powering the Portable Electronic Devices. *ACS Appl. Mater. Inter.* 2014. 7. 18225–18230.

[29] Jung, S.; Lee, J.; Hyeon, T.; Lee, M.; Kim, D.-H. Fabric-Based Integrated Energy Devices for Wearable Activity Monitors. *Adv. Mater.* 2014, 26, 6329-6334.

[30] Pu, X.; Li, L.; Song, H.; Du, C.; Zhao, Z.; Jiang, C.; Cao, G.; Hu, W.; Wang, Z. L. A Self-Charging Power Unit by Integration of a Textile Triboelectric Nanogenerator and a Flexible Lithium-Ion Battery for Wearable Electronics. *Adv. Mater.* 2015, 27, 2472-2478.

[31] Liu, L.; Pan, J.; Chen, P.; Zhang, J.; Yu, X.; Ding, X.; Wang, B.; Sun, X.; Peng, H. A Triboelectric Textile Templated by A Three-Dimensionally Penetrated Fabric. *J. Mater. Chem. A* 2016. 4, 6077-6083.

[32] Kim, K. N.; Chun, J.; Kim, J. W.; Lee, K. Y.; Park, J.-U.; Kim, S.-W.; Wang, Z. L.; Baik, J. M. Highly Stretchable 2D Fabrics for Wearable Triboelectric Nanogenerator under Harsh Environments. *ACS Nano* 2015. 9, 6394–6400.

[33] Zhong, J.; Zhang, Y.; Zhong, Q.; Hu, Q.; Hu, B.; Wang, Z. L.; Zhou, J. Fiber-Based Generator for Wearable Electronics and Mobile Medication. *ACS Nano* 2014, 8, 6273-6280.

[34] Zhong, J.; Zhong, Q.; Hu, Q.; Wu, N.; Li, W.; Wang, B.; Hu, B.; Zhou, J. Stretchable Self-Powered Fiber-Based Strain Sensor. *Adv. Funct. Mater.* 2015, 25, 1798-1803.

[35] Ishikawa, M.; Komi, P. V.; Grey, M. J.; Lepola, V.; Bruggemann, G. P. Muscle-Tendon Interaction and Elastic Energy Usage in Human Walking. J. Appl. Physiol. 2005, 99, 603-608.

[36] Starner, T. Human-Powered Wearable Computing. *IBM syst. J.* 1996, 35, 618-629.

[37] Qi, Y.; McAlpine, M. C. Nanotechnology-Enabled Flexible and Biocompatible Energy Harvesting. *Energy Environ. Sci.* 2010, 3, 1275-1285.

[38] Fan, F. R.; Tang, W.; Wang, Z. L. Flexible Nanogenerators for Energy Harvesting and Self-Powered Electronics. *Adv. Mater.* 2016, DOI: 10.1002/adma.201504299.

[39] Gao, Z.; Zhou, J.; Gu, Y.; Fei, P.; Hao, Y.; Bao, G.; Wang, Z. L. Effects of Piezoelectric Potential on the Transport Characteristics of Metal-ZnO Nanowire-Metal Field Effect Transistor. *J. Appl. Phys.* 2009, 105, 113707.

[40] Park, K.-I.; Son, J. H.; Hwang, G.-T.; Jeong, C. K.; Ryu, J.; Koo, M.; Choi, I.; Lee, S. H.; Byun, M.; Wang, Z. L.; *et al.* Highly-Efficient, Flexible Piezoelectric PZT Thin Film Nanogenerator on Plastic Substrates. *Adv. Mater.* 2014, 26, 2514-2520.

[41] Fan, F.-R.; Tian, Z.-Q.; Lin Wang, Z. Flexible Triboelectric Generator. *Nano Energy* 2012, 1, 328-334.

[42] McCarty, L. S.; Whitesides, G. M. Electrostatic Charging Due to Separation of Ions at Interfaces: Contact Electrification of Ionic Electrets. *Angew. Chem. Int. Edit.* 2008, 47, 2188-2207.

[43] https://en.wikipedia.org/wiki/Electrostatic_induction

[44] Wang, Z. L. Triboelectric Nanogenerators as New Energy Technology for Self-Powered Systems and as Active Mechanical and Chemical Sensors. *ACS Nano* 2013, 7, 9533-9557.

[45] Sessler, G. M. Electrets. 2nd, Berlin: Springer-Verlag (1987)

[46] Małecki, J. A. Linear Decay of Charge in Electrets. *Phys. Rev. B* 1999, 59, 9954-9960.

[47] Zhong, J.; Zhong, Q.; Fan, F.; Zhang, Y.; Wang, S.; Hu, B.; Wang, Z. L.; Zhou, J. Finger Typing Driven Triboelectric Nanogenerator and Its Use for Instantaneously Lighting up LEDs. *Nano Energy* 2013, 2, 491-497.

[48] Zhong, Q.; Zhong, J.; Hu, B.; Hu, Q.; Zhou, J.; Wang, Z. L. A Paper-Based Nanogenerator as A Power Source and Active Sensor. *Energy Environ. Sci.* 2013, 6, 1779-1784.

# CHAPTER ELEVEN

## Power Management in Wearable Body Area Networks

Cecília Lantos[1], and Haider Raad[2]

[1] *Simia Alba Research and Development Ltd., Budapest, Hungary*

[2] *Xavier Wearable Electronics Research Center (XWERC), Department of Physics, Xavier University, Cincinnati, Ohio, USA*

## Abstract

Wearable technology applied in health care initiates various new technical challenges. The site of physiological measurements is shifting from the clinical room to the patient's home and body area using continuous health monitoring schemes. Moreover, precise physiological measurements can be obtained in a real-time fashion but often at a lower resolution. To tackle such challenges, the sensor network needs to be carefully optimized through the design and configuration of the sensor nodes, in addition to their operation and communication strategies with the central node. The appropriate choice of the electronic elements for the sensor node including the power source, the sensor itself, the digital signal processing component, and transceiver(s) to build up an operating wearable body area network that consumes the lowest possible energy, is a cohesive task which will be the topic of discussion of this chapter.

**Keywords:** Wearable Body Area Network, Energy Management.

## 1.  Introduction

The application of wearable technology is growing in several areas such as smart homes, military and healthcare industry amongst others. Due to the rapid developments in microelectronics technology, flexible sensor materials [1], flexible electronics, miniaturization and manufacturing processes, wearable technological limitations are decreasing. However, there is still a lot of research needed to get a broader acceptability in many application areas. The challenges are various including the manufacturability, environment friendliness, industry planning, cost, security and privacy issues, in addition to the growing amount of data to analyze and process [2, 3].

Wearable technologies in health care applications bring new challenges since the measured physiological data in hospitals is usually measured via medical equipment installed in the hospital or a consulting room. Nowadays, the physiological data is becoming increasingly monitored at home, usually in or on the patient's body. Naturally, the common medical monitoring using wires is shifting toward a wireless connectivity which triggers new challenges. The communication protocol, appropriate radio frequency spectrum, and antenna design are amongst these challenges [3, 4].

The body area sensor network is built up from several sensor nodes which can be multimodal or dedicated to a particular physiological property. These properties could be either physical (heart rate, body motion, temperature, skin conductivity, swelling, blood pressure, brain activity, etc...) or chemical and biological (glucose, $O_2$, $CO_2$, etc...) [5, 6]. The measured signals are specific to the applications, with a purpose of diagnostics, wellness applications or therapeutic and assistive prevention [7].

The sensor types used to monitor physiological parameters can be worn, positioned on the body, or implanted. Special on body sensors are textile based such as yarns which are woven within a garment [8]. It is also worth mentioning that one of the newest technologies provides the possibility to implement nanosensors powered by the body itself [9].

## 2.   Body Area Network Energy Requirements

The body area sensor network requirements and restrictions differ from other network applications. The sensors collect, process, store and transmit data in real-time and continuously. The nodes communicate with the central node, and they can be connected and synchronized in an interoperative way; moreover, they are required to operate reliably while mobile [3-5].

The system requirements necessitate reliable communication between the sensing nodes and the central control unit. The central node can be a smartphone, PC or special device with a built-in microcontroller [3]. In body area sensor networks, the central processing unit (CPU) operates in an ultra-low power mode when there is no incoming information to be processed [10].

The network must be designed to be sustainable, scalable, secure, and safe for users [3]. The electromagnetic effects, interferences between wireless devices, and Specific Absorption Rate (SAR) must be taken into consideration during the design phase of such devices. To comply with wearable technology standards, specialized materials are often chosen for the sensor which must exhibit low profile, flexibility, and mechanical robustness [11]. It is also worth noting that the environment of the human body is especially challenging for the sensor operation due to its continuous physiological parameters variation which affects the sensing and actuation capacity [3, 4].

The circuitry is composed of a sensor, microprocessor and a transmitter (or a transceiver) capable of data acquisition, amplification, analog to digital (AD) conversion, data encoding, then controlling the wireless transmission by caching and packaging the data sent toward the central unit. The transmitted data is then stored in the computer for processing to be classified and displayed [4, 6]. Modern technologies such as System on Chip (SoC) are mostly utilized for the integration of the sensor network elements to reduce the size and cost of such implementations [3, 7].

The abovementioned scenario requires energy source and a memory to store the computed information to be transmitted wirelessly. Furthermore, such performance has to be optimized versus cost and power consumption [10]. This optimization is based on the choice of low-power hardware components, especially, the digital signal acquisition and processing unit [4].

The total energy of the body area network is split into two parts. One is for the circuitry, while the other is for the transmission. It should be noted that an optimization between the two power demanding tasks must be carefully planned [3].

Total energy can be saved based on the optimization of the transmission distance versus data rate. Also, battery-aware technology can help save energy, and the transmission time cannot be indefinitely long (deadline time) due to the real-time data collection. Also, the optimization procedure is different for the standard physical protocols and the modulation techniques [12]. Thus, the appropriate choice of the low-power hardware and the energy-efficient software (e.g.: the battery and digital processing algorithms) at the sensor node and the transmission network design will define the performance and power consumption of the wearable body area network [4, 9].

## 3. Battery Management and Energy Harvesting

The counterweight to the development of wearable wireless devices and sensors is the battery technology [7, 13]. The size, weight, conformability, lifetime of the sensor is substantially limited by the characteristics of the integrated battery.

When it comes to safety and biocompatibility, the material of the battery and its package is of paramount importance due to to the potential toxication hazard caused by its chemical contents [7].

On the other hand, the battery should be capable of supplying the required voltage for the sensor, microcontroller and transceiver for a reasonable duration. In addition to the continuous voltage supply, the maximum continuous current and pulse current capability define the effective battery capacity and lifetime [9].

The design of a low-power wireless network must take the battery size, energy storage capacity, and internal resistance into consideration. Using non-rechargeable (primary) batteries is a better fit for these applications considering their specifications: size, nominal voltage, maximum continuous current, and nominal capacity. The current research in battery technology is focused on overcoming the high power dissipation, minimizing the battery size and time between recharges, improving cumbersome sensor interface, and utilizing the maximum possible volumetric energy density [3, 7, 9].

Primary battery features are better compared to fuel cells since they exhibit higher energy densities, lower leakage rates, and lower initial cost. Examples of cells: CYC (cylindrical cell, $LiSOCl2$), BC (button cell, $Zn(OH)4$), and CC (coin cell, $LiMnO2$). Ultra\Super capacitors can also be used to power miniature electronics with a reasonably long lifetime. Their energy densities are typically lower than batteries but higher then normal capacitors [7, 9].

Lithium-ion polymer batteries (LiPo) are generally used in body area sensors although their size is a major drawback. On the other hand, Li-ion batteries are used as secondary (rechargeable) resources due to their electrochemical features [12].

Rechargeable (secondary) batteries play a vital role in lengthening the lifetime of sensors, but in some applications, they cannot be considered due to packaging limitations. Instead, energy harvesting methods can be a feasible option. Battery-less systems can be wirelessly powered or harvest energy from the environment. For example, RF (radiofrequency) energy harvesting from the environment is deployed through rectifying antennas (rectennas) [13, 14]. Other potential energy sources to be harvested are thermal energy such as body heat, temperature or pressure differences, and kinetic energy such as body motion, vibration, air or fluid flow, heart beat, and steps. Other sources include acoustic power, chemical/radioactive reactions, and ambient electromagnetic fields [7]. Furthermore, advancements in nanomaterials and molecular engineering will potentially expand the spectrum of energy harvesting elements. These alternative harvesters can possibly provide a solution to provide continuous energy supply in wearable wireless devices on the long run; however, their practical implementation is still under investigation.

## 4.  Data Processing at the Sensor Node

Clinical grade medical devices provide more precise measurements than wearables. The body environment can be hostile to their operation, and motion artifacts tend to degrade the sensitivity of the wireless sensors [3, 7]. In body area networks, wearable or implantable sensor nodes acquire, process and transmit data. The sensor drift and measurement mismatch degrade the signal integrity. This error can be avoided by the use of multimodal or redundant sensors, however, a trade-off arises as implementing more nodes is not recommended for body area networks. It is worth noting that advances in electronic microfabrication could serve as a remedy to this drawback [13]. On-node processing methods and drift correction computing can also improve the signal quality.

The data acquisition, processing, and transmission, can be enhanced by implementing processing algorithms at the sensor node. Processing data at the central unit (eg.: smartphone) is possible, but it can potentially compromise the privacy of the transmitted data.

If no complex computation needs to be implemented at the node, the acquired data can be stored for a small period of time before transmission, and a less powerful processor would be a more appropriate choice. Also, processing raw data using digital signal processing techniques at the sensor node can save energy since the reduced amount of data can be transmitted at a reduced data rate which consequently leads to a reduced transmission power [4, 5, 15].

Ideally, multiple nodes are synchronized to sample data using a dynamic sampling rate for transmission them to the central node to be processed. Typically, various sensor types can be used (RFID tags, implantable, wearable) while providing real-time data to the central node where signals from multiple sources are sampled synchronously [16]. The number of nodes is limited, but the system needs to operate securely with the operating nodes in case of error. Two to ten nodes are generally used in wireless body area network systems [15].

The sensors are configured to enable, disable, and dynamically set the sampling rate in an energy-efficient manner. On the other hand, admission control performs the triggering procedure, duty cycling mechanism, and periodically samples the signal. This leads to a reduced amount of data which saves power consumption while transmitting data. It should be noted that a lower duty cycle can compromise the performance when processing a vast amount of data (especially in imaging applications). This step is followed by data compression to improve the energy efficiency [17]. As proposed in [18], Compressed Sensing (CS) techniques are applied with Wireless Physical Layer Security (WPLS). The data sampling and encryption are implemented in a one-step algorithm to save power.

While in [10], a cross-layer framework has been developed based on unequal resource allocation to support biomedical data monitoring applications. In the developed framework, critical information (e.g.: ECG data) has been identified, and extra resources are allocated to ensure its reliable transmission.

The sensor node processing is a complicated computing task with resource-constrained power. To ensure a high signal quality, an energy-efficient medium access control (MAC layer) is necessary to support the various applications and data types [5, 10]. It is also worth noting that the sensor node processing lacks a common standard process. Such computing tasks are typically categorized as application specific, general-purpose, or domain-specific framework [5].

## 5. Transmission Networks

The communication protocol between the sensor nodes and the central unit provides the data transmission basis. The data is analyzed using classification tools at the central unit [4]. Since there are different distances involved in the transmission between the body area nodes and the central unit, and between the central unit and the consulting room, different communication protocol are deployed. The wireless transmission between the central unit and the control room is a long range transmission, so typically these standards are used: GSM, GPRS, UMTS, WLAN, WMTS. On the other hand, the communication between the sensor nodes and the central unit\node is a low-power short-range transmission wireless distance (Intra-BAN) [6]. IEEE S02.14.4(j) and IEEE S02.14.6 international standards offer reliable communication in body area networks ranging from narrow-band to ultra wideband in multiple low-power applications of implantable and wearable sensors. IEEE developed communication protocols for WBAN such as Bluetooth (802.15.1), low-energy light Bluetooth, and ZigBee (802.15.4). RFID standards are also used in addition to other protocols: MedRadio/Zarlink, Wireless Medical Telemetry System (WMTS), Sensium, ANT, WLAN 802.11a/b/g/n, IrDA, UWB [6, 7, 19-21].

The wearable and implantable sensors transmit data at a low rate, but the above-mentioned protocols support higher data rate. This gain gives the possibility to apply low duty-cycle hence saving power and reducing the actual bandwidth [7].

Operating at relatively high frequencies fulfills the requirement of consuming low transmission power at high data rate. The high-frequency energy-efficient communication platform needs to be optimized against the free-space data path loss [4, 21] since the path-loss in body area wireless networks is both distance and frequency dependent [7, 14]. At lower frequencies, however, the signal encounters less interference, and less Specific Absorption Rate (SAR) is exhibited in the body.

For implantable applications, lower frequency transmission is generally adopted since it is more suitable for lossy mediums. However, operating at low MHz frequency range gives rise to larger antennas. The classical trade-off arises here: low transmission rate, but longer transmission distance can be achieved [12, 22].

At high frequency, on the other hand, the body movements and environmental effects lead to a deterioration in the signal quality. This is attributed to tissue absorption, impedance mismatch, radiation pattern distortion, and fading. It is also important to remember that the low-power transmission and small antenna (low gain) lead to a reduced Signal to Noise Ratio (SNR), which consequently leads to a higher bit error rate [22].

A high-quality signal and reliable communication at an allocated transmission bandwidth must be supported by an energy-efficient MAC protocol. It is worth mentioning that a specific MAC protocol (for low-power wireless sensor networks) dedicated for health care application is still a challenging task [23].

Similar operations of wireless sensor networks trigger the need to implement a common network infrastructure for different applications [2]. This may help in implementing additional elements to the system to be configured for different users. However, the communication protocol for some specific applications may require different energy demands and carrier frequencies [10]. An adaptive communication protocol may provide a remedy to such compatibility issues [3].

Communication protocols secure data transmission by using security algorithms such as private key cryptography and data-encryption, which are resource-constrained. Security performance with a transmission bandwidth should be optimized against resource demands [5, 10]. Moreover, special attention must be taken to the security issues in wireless communication to overcome eavesdropping (privacy), impersonation (authentication), and jamming (maintainability) [24].

## 6.   Conclusion

The integration of wearable technology within healthcare applications can be further broadened by technological developments that fulfill the specific requirements of body area networks and wireless communication between the sensors and the central unit. Moreover, it is expected that this integration would be further revolutionized when the deployment of Internet of Things (IoT) is fully established [13].

The technological innovations are already addressing the practicality of this technology such as focusing on the development of miniature electronic components including sensors, batteries, microprocessors, and antennas. Furthermore, a dedicated transmission network protocol developed for body area communication, and a standardized digital processing protocol at the sensor node are already being seriously considered [7].

Wearable BAN systems should be supported by a reliable, low-power wireless network. Obviously, the power management limits the broader accessibility and usability of the targeted applications. Ideally, such systems should be sustainable and autonomous along its lifetime. This can be achieved by adopting an improved energy harnessing techniques. These possibilities are still under investigation [25].

# References

[1] Jeroen H. M. Bergmann, Salzitsa Anastasova-Ivanova, Irina Spulber, Vivek Gulati, Pantelis Georgiou, Alison McGregor, An Attachable Clothing Sensor System for Measuring Knee Joint Angles, *IEEE Sensors Journal*, Vol. 13, No. 10, 2013.

[2] Zhengguo Sheng, Chinmaya Mahapatra, Chunsheng Zhu, Victor C. M. Leung, Recent Advances in Industrial Wireless Sensor Networks Toward Efficient Management in IoT, *IEEE Access*, Vol. 3, 2015.

[3]Ashraf Darwish, Aboul Ella Hassanien, Review Wearable and Implantable Wireless Sensor Network Solutions for Healthcare Monitoring, *Sensors* Vol. 11 No. 6, pp 5561-5595. 2011.

[4] Subhas Chandra Mukhopadhyay, Wearable Sensors for Human Activity Monitoring: A Review, *IEEE Sensors Journal*, Vol. 15 (3), 2015.

[5] Giancarlo Fortino, Roberta Giannantonio, Raffaele Gravina, Philip Kuryloski, Roozbeh Jafari, Enabling Effective Programming and Flexible Management of Efficient Body Sensor Network Applications, *IEEE Transactions On Human-Machine Systems*, Vol. 43 (1), 2013.

[6] Ping Jack Soh,Guy A.E. Vandenbosch, Marco Mercuri, Dominique M.M.-P. Schreurs, Wearable Wireless Health Monitoring, *IEEE Microwave Magazine,* Vol.16 (4). 2015.

[7] Guang-Zhong Yang Editor, *Body sensor networks*, 2nd ed. Springer, 2014.

[8] Krisjanis Nesenbergs, Leo Selavo, Smart textiles for wearable sensor networks: review and early lessons, *IEEE International Symposium on Medical Measurements and Applications (MeMeA),* 2015.

[9] Edward Sazonov, Michael R. Neuman, *Wearable Sensors, Fundamentals, Implementation and Applications*, 1st ed. Academic Press, 2014.

[10] S. M. Riazul Islam, Daehan Kwak, Md. Humaun Kabir, Mahmud Hossain, Kyung-Sup Kwak, The Internet of Things for Health Care: A Comprehensive Survey, *IEEE Access*, Vol. 3, 2015.

[11] Ping Jack Soh, Guy A. E. Vandenbosch, Marco Mercuri, Dominique M. M. -P. Schreurs, Wearable Wireless Health

Monitoring: Current Developments, Challenges, and Future Trends, *IEEE Microwave Magazine*, Vol. 16 (4), 2015.

[12] Chenfu Yi, Lili Wang, and Ye Li, Energy Efficient Transmission Approach for WBAN Based on Threshold Distance, *IEEE Sensors Journal*, Vol. 15, No. 9, 2015.

[13] Benny P. L. Lo, Henry Ip, Guang-Zhong Yang, Transforming Health Care, Body sensor networks, wearables, and the Internet of Things, *IEEE Pulse*, January/February 2016.

[14] Paolo Nepa, Hendrik Rogier, Wearable Antennas for Off-Body Radio Links at VHF and UHF Bands: Challenges, the state of the art, and future trends below 1 GHz, *IEEE Antennas and Propagation Magazine* Vol. 57 (5), 2015.

[15] Min S. H. Aung, Faisal Alquaddoomi, Cheng-Kang Hsieh, Mashfiqui Rabbi, Longqi Yang, J. P. Pollak, Deborah Estrin, and Tanzeem Choudhury, Leveraging Multi-Modal Sensing for Mobile Health: a Case Review in Chronic Pain, *IEEE Journal of Selected Topics in Signal Processing* Vol. PP (99), 2016.

[16] Alex Page, Tolga Soyata, Jean-Philippe Couderc, Mehmet Aktas, Burak Kantarci, Silvana Andreescu, Visualization of Health Monitoring Data acquired from Distributed Sensors for Multiple Patients, *IEEE Global Communications Conference (GLOBECOM)*, 2015.

[17] Daniele Bortolotti, Bojan Milosevic, Andrea Bartolini, Elisabetta Farella and Luca Benini, Quantifying the Benefits of Compressed Sensing on a WBSN-based Real-Time Biosignal Monitor, *Design, Automation & Test in Europe Conference & Exhibition*, 2016.

[18] Ruslan Dautov, Gill R. Tsouri, Securing While Sampling in Wireless Body Area Networks With Application to Electrocardiography, *IEEE Journal Of Biomedical And Health Informatics*, Vol. 20, No. 1, 2016.

[19] Konstantina S. Nikita, *Handbook of Biomedical Telemetry*, 2014.

[20] Maulin Patel, Jianfeng Wang, Applications, Challenges, and Prospective in Emerging Body Area Networking Technologies, *IEEE Wireless Communications*, February 2010.

[21] Dolwin Ching Ching Kho, Ahmad 'Athif Mohd Faudzi, Dyah Ekashanti Octorina Dewi, Eko Supriyanto, Characteristics of Wireless

Technology for Healthcare Applications: An Overview, *IEEE Conference on Biomedical Engineering and Sciences*, 2014.

[22] Behailu Kibret, Assefa K. Teshome, Daniel T. H. Lai, Analysis of the Human Body as an Antenna for Wireless Implant Communication, *IEEE Transactions On Antennas And Propagation*, Vol. 64, No. 4, 2016.

[23] Muhammad Mahtab Alam, Elyes Ben Hamida1, Olivier Berder, Daniel Menard, And Olivier Sentieys, A Heuristic Self-Adaptive Medium Access Control for Resource-Constrained WBAN Systems, *IEEE Access*, Vol. 4, 2016.

[24] Z. Esat Ankarali, Qammer H. Abbasi, A.Fatih Demir, Erchin Serpedin, Khalid Qaraqe, Huseyin Arslan, A Comparative Review on the Wireless Implantable Medical Devices Privacy and Security, *EAI 4th International Conference on Wireless Mobile Communication and Healthcare (Mobihealth)*, 2014.

[25] Kiran Venugopal, Robert W. Heath, Jr., Millimeter Wave Networked Wearables in Dense Indoor Environments, *IEEE Access*, Special Section On Body Area Networks For Interdisciplinary Research Vol. 4, 2016.

Published by United Scholars Publications, USA
Copyright © 2016 United Scholars Publications
**www.unitedscholars.net**
**info@unitedscholars.net**

**ISBN-13:** 978-0692751718
**ISBN-10:** 0692751718

**Disclaimer**
The Publisher and the Editor\Author hold no liability for
incidental or consequential injuries or damages caused by the
information contained in this publication.

United Scholars
Publications

**Email: info@unitedscholars.net**
**www.unitedscholars.net**